"The AI Universe *is the conversation the* *Instead of fearmongering or hype, it offers grounded insight into how AI is reshaping every field and how we can adapt with wisdom. This is not just a book about technology; it's a book about the human present and future."*

—DAVID BROWNLEE, Bestselling Author of
Customer Service Success

"Erik Seversen has curated something extraordinary—an international dialogue about AI that bridges business, philosophy, and humanity itself. The multiple voices in this book don't just explain AI, they illuminate how humans can thrive alongside it. This book is a must-read for leaders, educators, and anyone curious about civilization's next leap."

—JARVIS LEVERSON, Founder of Accelerated
Growth Club

"The AI Universe *captures the global heartbeat of artificial intelligence including its challenges, its promise, and its profound impact on our shared destiny. With questions about whether AI should have a voice in the discussion of AI, this book elevates the conversation far beyond technology. It is brilliant, provocative, and deeply human."*

—Dr. MINA ANDERSON, AI Researcher and TEDx
Speaker

"The AI Universe *provides exactly what's missing from most AI conversations: nuance. It's not about utopia or apocalypse, it's about understanding. Each of the 34 experts in this book bring clarity to how AI is already transforming our industries, our thinking, and even our creativity. It is both insightful and timely.*"

—RUFUS GLAND, Director of Innovation, Global Institute

"*A fascinating blend of insight and inquiry. the contributors of* The AI Universe *have created a panoramic view of the AI revolution. It is practical, philosophical, and deeply inspiring. Whether you're an entrepreneur, policymaker, or lifelong learner, this book will help you see the bigger picture of what is happening with modern-day technology.*"

—SHAWN FECHTER, Senior VP of Sales, Author

"The AI Universe *invites us into a rare kind of discussion—one that honors both human imagination and machine intelligence. It's a thoughtful roadmap for thriving, not just surviving, amid the most transformative era since the internet. Read it, reflect on it, and join the conversation.*"

—CASEY LINDBERG, SpaceX, Quality Control Manager

THE AI
UNIVERSE

THE AI UNIVERSE

Thriving Within Civilization's Next Big Disruption

Authored by:
Erik Seversen, Gavin W H Anderson, Gabriela Bar, PhD,
Christopher Bosley, LingYi Chang, Naser Chowdhury, PhD,
Will Conaway, Kunal Devidasani, Astrit Dibrani, Veda Duman
Kantarcıoğlu, PhD, Leonardo Camargo-Forero, PhD, Jian
Guo, SamDavid Jeyaraj, Nainish Kapadia, Swaroop Kariath,
Nikolaos Lampropoulos, Anastassia Lauterbach, PhD,
Bettina S. Lippisch, Marion Løken, PhD, Colin Mansell, Nithin
Mathews, PhD, Patrick J. Meyers, Ángel Moyano, Giorgio
Natili, Mohamed Omer, Pramod M. Patke, Robert Pluska,
Niyati Prajapati, Vasanthan Ramakrishnan, Ruan Schutte,
Sheily Sharma, Sakina Syed, Hanoz Tabak, Daniel Jonathan
Valik, Hardik Modi, Jan Wiersma

THIN LEAF PRESS | LOS ANGELES

The AI Universe: Thriving Within Civilization's Next Big Disruption individual chapters. Copyright © 2025 by Gavin W H Anderson, Gabriela Bar, PhD, Christopher Bosley, LingYi Chang, Naser Chowdhury, PhD, Will Conaway, Kunal Devidasani, Astrit Dibrani, Veda Duman Kantarcıoğlu, PhD, Leonardo Camargo-Forero, PhD, Jian Guo, SamDavid Jeyaraj, Nainish Kapadia, Swaroop Kariath, Nikolaos Lampropoulos, Anastassia Lauterbach, PhD, Bettina S. Lippisch, Marion Løken, PhD, Colin Mansell, Nithin Mathews, PhD, Patrick J. Meyers, Ángel Moyano, Giorgio Natili, Mohamed Omer, Pramod M. Patke, Robert Pluska, Niyati Prajapati, Vasanthan Ramakrishnan, Ruan Schutte, Sheily Sharma, Sakina Syed, Hanoz Tabak, Daniel Jonathan Valik, Hardik Modi, Jan Wiersma

Library of Congress Cataloging-in-Publication Data
Names: Seversen, Erik, Author, et al.
Title: *The AI Universe: Thriving Within Civilization's Next Big Disruption*
LCCN: On File

ISBN 978-1-968318-18-5 (hardcover) | 978-1-968318-17-8 (paperback)
ISBN 978-1-968318-16-1 (eBook) | 978-1-968318-19-2 (audiobook)

Artificial Intelligence, Science & Technology, Business, Professional Development
Cover Design: 100 Covers
Interior Design: Dindo Sanguenza
Editor: Nancy Pile
Thin Leaf Press
Los Angeles

THIN
LEAF

Thank you for reading this book. There is information found within the following pages that can greatly benefit your life, but don't stop there. Make sure you get the most you can from this book and reach out directly to the expert-authors who want to help you to use AI to your advantage, to thrive within civilization's next big disruption, and to manifest success in your life. Contact information for each author is found at the end of their respective chapter.

To the innovators who are expanding the horizons
intelligence in our universe.

CONTENTS

INTRODUCTION xix
By Erik Seversen
Author of *Ordinary to Extraordinary*
Los Angeles, California

CHAPTER 1 1
AGENTIC AI: YOUR NEW BEST FRIEND?
By Gavin W H Anderson
AI Ideas and Visions for People and the Planet
Buckingham, England, United Kingdom

CHAPTER 2 9
A DYSTOPIAN FUTURE IS NOT INEVITABLE
By Gabriela Bar, PhD
AI Law and Ethics Specialist, Attorney at Law
Wrocław, Poland

CHAPTER 3
AI ISN'T REPLACING YOU—BUT SOMEONE WHO
USES IT WILL 19
By Christopher Bosley, MLS
Machine Learning and AI Consultant
Salt Lake City, Utah

CHAPTER 4 27
BETTER TOGETHER: WHY AI'S PROMISE
REQUIRES HUMAN CONNECTION
By LingYi Chang
AI Consultant and Speaker, Co-Founder of illumi.one
Singapore

CHAPTER 5 39
MARKETPLACE FOR INTELLIGENCE: AN
ESSENTIAL DIFFERENTIATOR FOR INTELLIGENT
ENTERPRISES AND INDIVIDUALS
By Naser Chowdhury, PhD
Applied AI and ML Practitioner; Technology Architect
Phoenix, Arizona

CHAPTER 6 51
REIMAGINING HEALTHCARE: A
TRANSFORMATIONAL CALL TO ACTION
By Will Conaway
AI Advisor, Healthcare Strategist, Educator
Charlotte, North Carolina

CHAPTER 7 65
AUTOMATION UNLEASHED: TRANSFORMING
BUSINESS OPERATIONS WITH AI AND ROBOTICS
By Kunal Devidasani
Tech Consultant—AI & RPA
Jersey City, New Jersey

CHAPTER 8 75
THE AI UNIVERSE: THRIVING IN A WORLD WITH
AI, SMART MACHINES, AND ROBOTS
By Astrit Dibrani
Founder & CEO CloudAstro; Serial Entrepreneur
Munich, Germany

CHAPTER 9 81
INTELLIGENT SAFEGUARDS: AI AT THE
FRONTLINES OF RISK, RESILIENCE, AND RESPONSE
By Veda Duman Kantarcıoğlu, PhD
Nuclear Engineer, Expert in Defense Industries and DRM
Ankara, Türkiye

CHAPTER 10 91
AWAKENING BEYOND THE SINGULARITY
By Leonardo Camargo-Forero, PhD
CEO, UbiHPC. Science-fiction Author, *The Dark Buddha*
Bucaramanga, Colombia

CHAPTER 11 101
STRATEGIC ALCHEMY: TURNING DATA INTO
ENTERPRISE GOLD WITH AI LEADERSHIP
By Jian Guo
AI Executive and Practitioner
Warren, New Jersey

CHAPTER 12 111
THE SILENT CATALYST: AI'S RISE, REACH, AND
THE HUMAN RECKONING
By SamDavid Jeyaraj
Strategic Tech Leader; Scaling AI
Brussels, Belgium

CHAPTER 13 121
AI-DRIVEN ENTERPRISES: NAVIGATING CHANGE
IN THE AGE OF GENERATIVE INTELLIGENCE
By Nainish Kapadia
Product Management, Gen AI
Bengaluru, India

CHAPTER 14 131
POWERING THE NEXT ENERGY REVOLUTION
By Swaroop Kariath
Energy Industry Leader and AI Advocate
Calgary, Canada

CHAPTER 15 147
INTELLIGENCE AMPLIFIED: HOW AI
TRANSFORMS BRAND NARRATIVES AND GROWTH
By Nikolaos Lampropoulos
Founder of Shapes + Numbers, AI Advisor
New York, New York

CHAPTER 16 159
AI LITERACY AND THE NEXT GENERATION OF
DIGITAL THINKERS
By Anastassia Lauterbach, PhD
CEO AI Edutainment, Professor for AI
Basel, Switzerland

CHAPTER 17 177
FROM BLACK BOX TO GLASS BOX: BUILDING
TRUST AND PROTECTING PERSONAL DATA IN
THE AGE OF AI
By Bettina S. Lippisch
Intelligent Enterprise Transformation, AI Governance
Chicago, Illinois

CHAPTER 18 191
YOU TOO, SHOULD BECOME AI-AUGMENTED—
HUMAN-CENTERED AI FOR A SUSTAINABLE FUTURE
By Marion Løken, PhD
Digitalization Leader; Top 50 Women in Tech
Oslo, Norway

CHAPTER 19 203
LEARNING IN THE AGE OF AI: WHAT MACHINES
CAN'T TEACH US
By Colin Mansell
CEO and Founder, Skills U
Singapore

CHAPTER 20 217
BLUEPRINT FOR TOMORROW: DIGITAL
STRATEGY FOR FINANCIAL SERVICES
By Nithin Mathews, PhD
AI Researcher, Financial Industry Advisor, Startup Mentor
Zurich, Switzerland

CHAPTER 21 229
THE GRID THAT THINKS
By Patrick J. Meyers
Energy Executive—LC Energy Partners; Author: *Tariff Nation*
Houston, Texas

CHAPTER 22 239
FROM TECHNOLOGY TO TRANSFORMATION: AI
AS A CEO'S STRATEGIC LEVER
By Ángel Moyano
AI Transformation Strategist
Madrid, Spain

CHAPTER 23 251
BUILDING TRUST IN AN AGENT-DRIVEN WORLD
By Giorgio Natili
Head of Engineering
Seattle, Washington

CHAPTER 24 267
ORCHESTRATING TRANSFORMATION:
STRATEGIC LEADERSHIP AT THE INTERSECTION
OF AI, GOVERNANCE, AND CUSTOMER EXPERIENCE
By Mohamed Omer
AI Advocate
Frederick, Maryland

CHAPTER 25 277
DIGITAL DARWINISM: THE NEW RULES OF
SURVIVAL IN AN AI WORLD
By Pramod M. Patke
Product Manager, System Architect
Gothenburg, Sweden

CHAPTER 26 291
INDUSTRIAL AI UNLOCKS HIDDEN POTENTIAL
By Robert Pluska
Industrial AI Strategist, Advocate, and Founder of Advanced Solutions
at JSP
Dusseldorf, Germany

CHAPTER 27 303
A ROAD TO AGENCY: FROM MULTIMODALITY TO
SUPERINTELLIGENCE
By Niyati Prajapati
AI Lead and Tech Executive
Mountain View, California

CHAPTER 28 311
HOW TO DESIGN GOVERNANCE THAT ENABLES,
NOT BLOCKS
By Vasanthan Ramakrishnan
Entrepreneur, Legal & Compliance Expert and Angel Investor
Chicago, Illinois

CHAPTER 29 333
THE INTENTIONAL CYBORG: HUMAN UPGRADE
REQUIRED
By Ruan Schutte
Innovation and AI Projects Specialist
Johannesburg, South Africa

CHAPTER 30 345
HUMAN-CENTERED AUTOMATION: DESIGNING
WITH CONSCIENCE, BUILDING WITH INTENTION
By Sheily Sharma
Architect, Intelligent Systems; Advocate, Ethical Automation
London, England, United Kingdom

CHAPTER 31 357
AI IN ROBOTICS: REINFORCEMENT LEARNING
FOR ROBOTIC CONTROL
By Sakina Syed, B.Sc.
Microsoft IT Professional, AI Engineer and Consultant
Toronto, Canada

CHAPTER 32 371
MARKETING AND AI—HOW MARKETERS CAN
USE AI TO ACHIEVE THEIR GOALS
By Hanoz Tabak
Strategic Marketing Leader
Toronto, Canada

CHAPTER 33 379
THE DAWN OF CLOUD-NATIVE CUSTOMER
ENGAGEMENT: PERSPECTIVES ON CCAAS AND
THE AI-DRIVEN REVOLUTION
By Daniel Jonathan Valik
VP Product Management, CCaaS and AI Services
Seattle, Washington
By Hardik Modi
Director Product Management, CCaaS Services
San Francisco, California

CHAPTER 34 395
ALCHEMY IN THE AI ERA: THE MINDSET AND
CULTURE THAT TURNS FAILURE INTO INNOVATION
By Jan Wiersma
Entrepreneur, Technologist, and Venture Partner
Amsterdam, Netherlands

DID YOU ENJOY THIS BOOK?...................................405

INTRODUCTION

By Erik Seversen
Author of *Ordinary to Extraordinary*
Los Angeles, California

> *We're not just building machines; we're building our future.*
> —Max Tegmark

What does it mean to have a voice? I commonly hear statements like, "My voice is not being heard," "The XYZ group's voice is not being represented in this dialogue," "If only I could find my authentic voice…," or "Everyone in this discussion should have a voice." Of course, the subtlety here is that a voice is more than the sounds coming from the mouth. It is having representation in a conversation. My initial question about this, for this book about AI, is—who should have a voice in the discussion about AI as civilization's next big disruption?

One of the goals of this book is to gather the voices of AI experts from around the world to share their ideas about AI and how humans can thrive in this new era. As I begin to think about these ideas, I wonder, does an AI's voice count? Is AI capable of adding value to a discussion? Is an AI's response to a prompt just a rearrangement of millions of human thoughts (bits of information) found in the large language model the AI has been trained on? Really, my question is— does or should AI have a voice in human discussions about important things? Following this train of thought, more questions arise: Is AI more like a calculator whose function is to speed up mathematical calculations, allowing us to arrive at mathematical conclusions more quickly or is AI beyond that? Can AI influence not only the speed of thoughts, but the thoughts themselves? If an AI responds to a prompt with a unique perspective about something (gathered from a selection of prior human perspectives), is that an original idea? What if the AI response sparks an absolutely unique and original idea within the human who wrote the prompt and read the response? Can we say the root of that totally original thought is human or AI-inspired?

While I have opinions about some of the questions above, I really don't know how AI is going to fit within human philosophy and human evolution, but one thing is clear. The AI box has been opened, and human civilization will never be exactly the same. Whether you like it or not, AI is here.

As I mentioned, the goal of this book is to bring together a number of AI experts to shed some light on what is happening with AI in various industries. You'll get to those ideas in the book, but since I use ChatGPT so often, like I have done in previous books, I want to again allow Alvin (the name I've assigned to my AI) to share his voice on the topic of this book, AI as civilization's next big disruption. With that in mind, I asked my ChatGPT if AI should have a voice in the discussion about AI's disruption within human civilization. Here is the response:

At first glance, it may sound strange to suggest that AI should have a voice in its own story. After all, AI does not have consciousness, values, or lived experience. Its "voice" is ultimately a reflection of the human data, structures, and intentions behind it. Yet ignoring AI's output entirely would also be shortsighted. AI systems are already producing insights, generating patterns across data too vast for any human to process, and offering novel perspectives on science, language, and art. In this sense, AI can contribute to the conversation—not as an independent moral agent, but as an amplifier of human imagination and an accelerant of human inquiry.

The key distinction is this: AI can inform the discussion, but it cannot own it. The responsibility for interpretation, ethics, and decision-making rests squarely with humanity. The danger lies not in giving AI a voice, but in mistaking its outputs for wisdom. Wisdom is born of lived experience, suffering, empathy, and moral reflection—qualities AI does not possess.

Therefore, AI's role in the great conversation about civilization is similar to a brilliant tool or collaborator: It can surface possibilities, expand our thinking, and even challenge our assumptions. But the responsibility for meaning, values, and choices belongs to us. Humanity must remain the author, even as AI helps hold the pen. In the end, the story of AI will not be written by machines, but by us—together, with wisdom and care.

Well, there you have it. ChatGPT suggests that humanity must remain the author of civilization… and I'm slightly perplexed by the usage of the possessive determiner "our" and the object pronoun "us" in the last paragraph. While AI might confess to not having lived experiences and wisdom, it certainly seems to want to mimic these traits as it continually sees them in its human-developed models. And, today's AI is still really in its stage of infancy. As AI evolves, it will be very interesting to see how blurred the line becomes as definitions of machine "consciousness" and "self" become real discussions.

While I'm fascinated by these ideas, the main point of this book isn't to locate our new reality within the nexus of human-machine interaction, but to show readers how AI is being used in various industries right now, what changes might be occurring in our lives and our workplaces, and how humans can thrive within this new era of AI.

Since I'm not an AI expert, I didn't try to write this book about AI by myself. Rather, I solicited the help of 34 AI experts from various backgrounds and locations. The co-authors of this book come from all over the USA, Canada, the United Kingdom, Norway, Sweden, Belgium, the Netherlands, Germany, Poland, Switzerland, Spain, Turkey, Colombia, India, Singapore, and South Africa. These authors have been featured on NBC, FOX, ABC, CBS, BBC, CSUN, Sky News, and more. They are professionals who are university professors, business owners, consultants, engineers, software developers, lawyers, strategists, marketers, cybersecurity leaders, healthcare experts, advisors, TEDx and keynote speakers, researchers, senior data analysts, IT directors, CTOs, and more. The one thing these individuals have in common is that they all have something to share about artificial intelligence, and these ideas are available to you now.

Although this book is organized around the united theme of thriving in the new era of AI, each of the chapters is totally stand-alone. The chapters in the book can be read in any order. I encourage you to look through the table of contents and begin wherever you want. However, I urge you to read all the chapters because, as a whole, they provide a great array of perspectives. Each is valuable in helping you understand how AI is being used in many areas of human industry.

It is my hope that you discover something in this book that helps you navigate civilization's next big disruption as humans and machines become more interdependent and AI technology continues to insert itself into many aspects of our everyday lives. It is my hope that you adapt well within our transforming universe as AI begins to impact more and more areas of our lives.

Whether AI is the best thing to happen to humans as a species or the start of our downfall, AI is here, and it is my hope that we are able to work with machines to begin creating a lifestyle both with greater productivity and more time for activities or with family and friends. Like never before, humans and machines, working together, have the ability to determine our destiny. With voices from academic and corporate professionals to technicians and engineers to poets and artists and to AI itself, let's enjoy this important time and hopefully set the stage for a great future.

About the Author

Erik Seversen is on a mission to inspire people. He holds a master's degree in anthropology and is a certified practitioner of neuro-linguistic programming. Erik draws from his years of teaching at the university level and years of real-life experience in business to motivate people to take action, creating extreme success in business and in life.

Erik is a TEDx and keynote speaker who has reached over one million people through his public speaking and live courses. He has visited 99 countries and all 50 states in the USA and has climbed the highest mountains on four continents, 15 countries, and 18 states. Erik has published 19 bestselling books on the topics of mindset, success, and peak performance, and he has helped over 400 people become authors. He is a full-time writer, book consultant, and speaker, and he lives by the idea that success is available to everyone—that living an extraordinary life is a choice.

Erik lives in Los Angeles with his wife and has two boys currently studying at university.

Contact Erik for interviews, speaking, or book publishing consultation.

Email: Erik@ErikSeversen.com

Website: www.ErikSeversen.com

LinkedIn: https://www.linkedin.com/in/erikseversen/

CHAPTER 1

AGENTIC AI: YOUR NEW BEST FRIEND?

By Gavin W H Anderson
AI Ideas and Visions for People and the Planet
Buckingham, England, United Kingdom

> *AI will probably be smarter than any single human next year. By 2029, AI is probably smarter than all humans combined.*
>
> —Elon Musk *(posted on X)*

The Future

Welcome to a glimpse of the future. You've heard a lot about AI taking over the world; it may not take over, but it will change it fundamentally, just as the wheel and the internet have. Possibly delivering even more of a shift, as it will impact how we live, work, learn, and earn, so AI agents or agentic AIs are the new revolution. I'll explain the difference between an AI agent and agentic AI in a moment, but first, we should provide some context around what AI means to us, as a human race, in general and how quickly or slowly the changes will occur.

Robots for Breakfast

So, yes, AI will bring humanoid robots, but not for a long while. Far sooner, we will see simple $300 robotic arms at the local deli or coffee shop before we meet a robot walking down the street. The same is true in our factories currently. We will have exoskeletons (a form of robotics) that support human workers way before humanoid robots. Ford has been a pioneer in using exoskeletons, integrating them into its assembly lines in 15 factories globally. These exosuits help workers lift and manoeuvre parts that might otherwise be too heavy, aiming to reduce work-related illnesses like musculoskeletal disorders or RSI (repetitive strain injuries). Full humanoid robots are a long way away from our streets and even further away from taking over our world. Robots are not the current revolution; AI is. The word "revolution" is apt, albeit a slow one over the next few years, but a fundamental shift like humanity has never seen before.

Who Are the Agents?

Agents or personal AI have been mentioned by Meta recently, with Mark Zuckerberg referring to them as your new "Facebook friend," but agents are far more than that. Also, why are they not mentioned more realistically? As helpers and personal champions, this is the first impact you, your children, and your grandparents will experience from the AI revolution.

Let's return to the difference between an "AI agent" and an "agentic AI". An AI agent is your voice-activated AI. Consider JARVIS (Just A Rather Very Intelligent System) from *Iron Man*, who always supports but never takes over and is always subservient, yet quite possibly more intelligent or well-versed than Iron Man or Tony Stark—a perfect companion. Tony Stark in *The Avengers: Age of Ultron* shows emotion when he thinks his AI friend JARVIS has just been killed, so a relationship was built. Is this science fiction? A genuine emotional connection and humans showing empathy to AI will be real; to those who decry it, Ihave one word: Tamagotchi.

To summarise the difference, an AI agent can execute tasks based on the commands it's given. An agentic AI can manage multiple agents until it completes a task.

An AI agent is the worker concentrating on a single or specialised task.

Agentic AI is a manager having access to multiple agents to complete tasks for them.

Our personal AIs will actually be agentic AIs because they will be able to make decisions for us, follow through, and complete an entire task using multiple agents. What could possibly go wrong?

What If They Did Good?

We have a current fear of AI taking over, where AI can sell us things, possibly give us incorrect advice, or be too persuasive, take our jobs, or know the exact product we want at that perfect moment and know how we can be convinced. But what if we had one of these agents that were actually "on our side" or a filter/helper, a personal agent that can help you, keep you secure, and tell you what is fake, what is real, what is an absolute bargain, and what is a scam? We are hearing lots of rhetoric about what AI will do to us, but what about what AI can do for us?

The world will refer to them as AI agents, so, for simplicity, we will also refer to them as personal agents. Let's call our personal agent Alan (our JARVIS).

What will the personal agent be able to do? We must begin by examining the current state of AI agents and personal agents. The improvement is actually hampered partly by hardware restrictions and processing power. A personal agent would run on a local instance or a local PC, let's say, in the corner of your home or, once handheld power is available, on your phone, and connect to the internet. The important part about an agent is that it stands away from all the other AIs, so it can actually handle your personal information and reach out to utilise AI agents such as ChatGPT, DeepSeek, Dali, or any other

AI you require. You'd then pay a monthly subscription for these AI services. Still, your own agentic AI Alan will sit at the core.

So, how will Alan actually manifest himself in our lives? He will be available by a voice prompt on our phone, home PC, or even through our IoT, but security will always be a concern. Alan will be able to actually deal with all your personal needs as well as being your personal advisor, your educator, your mentor, and, of course, your financial advisor, hence the requirement for security. Alan will be your PA and will organise meetings, book the car for service, and remember birthdays and anniversaries. However, Alan will be far more than just a glorified diary or auto-reminder tool. It will be able to teach your children specialised subjects; Alan will also serve as your translator when you're on holiday or teach you Spanish before you depart. You probably booked that holiday through Alan, as he knows your preferences and searches out the correct flights and accommodations that are tailored to your personal taste.

That's the simple stuff. What Alan will actually be able to do is utilise agentic AI, and ideas will be set free. What do I mean by that? You have probably had a business idea, but the legal aspects, setting up, marketing strategy, and customer support are areas where you may lack expertise. But with agentic AI, this will be very different. Imagine if you wanted to start an online business, such as one specialising in car cleaning products. You could tell Alan what your target market was, the products you want to sell, where to look for them, possibly source suppliers from China, design a brand that appeals to your target market, create the website, plug in a Shopify site or similar, create the agreement with the Chinese supplier, and literally, you could actually be sitting on the beach while Alan is doing all this for you. This will also involve a shift from being employed to working for yourself or a friend who has set up a new entity. So, yes, AI will disrupt the employment market, but it will also provide opportunities we have not even thought of yet. AI will also bring about a significant reset; some of you may recall when computing first began, and we all discussed working three-day weeks due to the technology; well, that may be part of your children's future.

Master or Mentor?

If this is getting scary, let's take a step back and ground ourselves with some simple facts: AI can only utilise human knowledge; there is no other knowledge beyond that. Additionally, current-moment processing speeds are limiting public agentic AI to a slow process, with a maximum processing speed of approximately 30 minutes to one hour. With quantum computing on the horizon and hardware improvements in processing and memory with instant access, allowing us to move faster and faster, the agentic model should start emerging soon, I would estimate, by 2028.

Today, we have the equivalent of having a Nokia with an LCD screen, and we are trying to imagine the iPhone. That is where we are with AI; we are staring in wonder but can't see how big a shift this will be.

This new world of personal agents will cause chaos in the same way the Industrial Revolution did. However, with the turmoil and development, providing every single person on earth with access to a personal agent so that they can be educated, supported, helped, and advised can only be a benefit to humanity. Yes, there will be a difference between the haves and the have-nots, so who can actually access agents or afford to pay the subscription for agents, but that's not the point. The point is that this represents a fundamental shift in how we learn, engage, and conduct business.   Currently,   not everyone can write prompts, but when AI becomes a handheld voice-activated personal agent, that will be a game changer, even for your grandparents, whom I mentioned earlier. They will be able to actually talk to somebody, possibly the voice of their favourite gardening adviser, off the radio. They'll be able to receive advice on how to care for their garden, plant vegetables, and address physical care and medical needs, with the option of having a personal doctor on site to monitor them 24/7. Finally, we will unlock the internet for those who find keyboards and web browsers a barrier.

We have a lot to work out about AI before we dive into this new world, but it's coming at us like a runaway train! I think my advice would be to teach your children how to use AI. Not all AI will be beneficial; yes, we will likely see criminal or illegal activities. Remember when Facebook started? It was a fantastic place when it began, doing infinite good by reconnecting old friends. However, Facebook eventually turned into an advertising and influencing network that may not be purely driven by good today.

AI Security

Tim Berners-Lee, the inventor of the World Wide Web, had expressed concerns years ago about the security implications of the web. He was advocating for a decentralised and user-controlled internet, where individuals had more agency over their personal data, much like having a postal agent. So far from personal agents being the forerunners of Skynet and Terminator, they will actually be our personal assistants, coaches, mentors, and friends, but they won't be able to show empathy. Still, indeed, they will be able to organise people's lives and also help them with exercise and diet prompts, suggesting what they should be eating. All this is obviously very optional, but with AI, it can actually help us become the best version of ourselves. So, is this a brave new world, just what we were promised with the internet?

The Future Is Bright

Some big thoughts, but we will all be able to harness all human knowledge (the internet) and have a personal assistant and guide (agentic AI) delivered through IoT in a decentralised, omnipresent way (yes, I do want Tony Stark glasses).

AI, in summary, will be as significant a change as the advent of flight, the invention of cars, or the emergence of the internet. Embrace the decentralised age of intelligence; the agentic AI will be on your side.

About the Author

Gavin Anderson's passion for technology took him through the digital revolution in music, video, and special effects. Creating world events for Apple and Microsoft, after forming his first startup in the early 2000s, he never went back to the mainstream. Now Gavin is a chair of a global streaming tech company (Aitum.tv) and chair of AI neurology memory company (Memorifytech.com). He is also writing and consulting in the area of digital transformation, both at the system and product levels.

Email: gavin@n2collective.com

Website: www.n2collective.com

LinkedIn: https://www.linkedin.com/in/gavinwhanderson/

CHAPTER 2

A DYSTOPIAN FUTURE IS NOT INEVITABLE

By Gabriela Bar, PhD
AI Law and Ethics Specialist, Attorney at Law
Wrocław, Poland

> *If machines are to be placed in a position of being stronger, faster, more trusted, or smarter than humans, then the discipline of machine ethics must commit itself to seeking human-superior (not just human-equivalent) niceness.*
>
> —Nick Bostrom, Eliezer Yudkowsky, 2011

Defining Human Identity

Speaking of artificial intelligence is, unavoidably, speaking of ourselves. The emergence of intelligent machines has reopened an ancient and still unresolved question: What does it mean to be human? The boundary between human and machine, once assumed to be self-evident, is now deeply unstable. As we increasingly integrate technology into our minds and bodies—and as machines begin to mirror human behaviors, judgments, and even emotions—the very essence of human identity becomes contested ground.

What then still distinguishes us from AI? Is it consciousness, moral intuition, emotions, or a sense of self? These questions—once relegated to metaphysics—have become practical and urgent. We live in a world where prosthetic limbs, neural implants, and digital assistants blur the lines between biology and code. Is a person still human if their thoughts are preserved on silicon, if their body is mostly mechanical, if their decisions are shaped by algorithms? These are not future hypotheticals. They are the edge cases of our present.

Yet even as these boundaries blur, no scientific or commonsense consensus exists on definitions of consciousness, intuition, or even emotion. The European AI Act, which regulates emotion-recognition systems, first had to stipulate what does not count as emotion—excluding purely physical states such as pain or fatigue. In other words, our legal frameworks are already forced to legislate definitions that philosophers have never settled, underscoring how fragile the conceptual ground beneath this debate remains.

We are witnessing concrete efforts to digitally replicate human consciousness and cognition. The Human Brain Project, for instance, invested over a decade of multidisciplinary research into developing detailed digital brain models and sophisticated simulations of neural networks—though full brain emulation remains an aspirational goal rather than a realised achievement. Meanwhile, initiatives like the Brain Preservation Foundation and Alcor Life Extension Foundation actively preserve human brains and bodies, betting on future technologies capable of cognitive resurrection or reanimation.

This is no longer mere science fiction. It is an emerging frontier of transhumanist ambition quietly shaping our conception of life and consciousness itself. Transhumanism, at its core, is the belief that human limitations, including death, can and should be overcome through technology. Whether through brain emulation, cryonic preservation, or neural enhancement, the underlying promise is radical: Human identity is not sacred or fixed, but hackable, upgradeable, and potentially eternal.

In one of his novels, the renowned science-fiction author Stanisław Lem imagined disembodied brains kept alive in sensory isolation—consciousness severed from the body and nourished only by prerecorded stimuli. This chilling scenario foreshadowed contemporary discussions on mind-uploading, brain preservation, and simulated reality. Lem foresaw not only the technical possibility but also the existential and ethical unease: Are these isolated brains truly alive? Do they count as persons? Do they experience suffering? More unsettling still—how different are we, increasingly dependent on algorithms and enclosed within the carefully curated artificial realities of social media, metaverses, and digital avatars?

Liberal traditions have long celebrated the idea of the autonomous individual—free, rational, and self-determined—but this vision grows fragile under scrutiny. Neuroscience, psychology, and behavioral economics have all shown that human agency is profoundly constrained by biology, history, and context. We are not as free as we think. We do not consciously choose our desires, our tendencies, or our emotional responses. Even our most personal decisions—what we value, whom we trust—are shaped by forces beyond our conscious control, and one of these forces are Big Tech algorithms. Social media subtly directs our attention, amplifies emotions, and influences political beliefs. Dating apps match us according to hidden, profit-driven criteria, while streaming services guide our cultural tastes by carefully curating what we watch. Each click, swipe, or like is nudged by invisible but powerful influences, engineered by the tech industry to shape human behavior beneath our conscious awareness.

In this context, our anxiety about AI autonomy begins to look strangely misplaced. Why do we fear autonomous machines when we barely understand our own autonomy? Why do we demand full transparency from AI, when our own minds are often opaque, even to ourselves? The truth is unsettling: We fear AI not because it is radically other, but because it is uncannily familiar. Its intelligence mirrors ours: flawed, contingent, programmable.

Nevertheless, it is precisely this resemblance that holds the greatest promise. If AI compels us to reconsider what it means to be human, we must not shy away in fear. We must, instead, confront the ambiguity. Engaging with AI is like holding a mirror up to our own cognitive and moral architecture. In doing so, we might begin to see ourselves not as finished beings, but as evolving, imperfect, and deeply entangled with technology. The ethical challenge, then, is not merely about building safe AI. It is about redefining our values in response to it. This demands humility, curiosity, and an openness to rethinking what we value in ourselves.

Protecting Mental Integrity

Advances in brain research have moved from digitally simulating brain processes towards technologies that interact directly with neural tissue. Neural implants are already in clinical use, effectively restoring vision, controlling tremors associated with diseases like Parkinson's, and enabling communication for individuals with severe motor impairments. However, the potential to restore neural functions also inherently includes the potential to alter them, raising significant concerns regarding mental integrity, once implicitly protected by biological limits, now directly accessible through technology.

Artificial intelligence amplifies these concerns, as sophisticated machine-learning models become capable of recording, classifying, and predicting neuronal activity with remarkable precision. Such AI systems, in combination with neuroimaging techniques like functional magnetic resonance imaging (fMRI), can already reconstruct visual images that subjects have seen or even imagined. Research initially conducted in controlled clinical environments is now extending into consumer neurotechnology, such as wearable headbands or helmets marketed for improving concentration and mood, or inducing meditative states or lucid dreams. These consumer devices collect neural data at scale, typically outside regulated medical contexts.

The imbalance of power is evident: Technology corporations control the development and deployment of these neurotechnologies, whereas end users frequently surrender sensitive neural data with minimal awareness or protection. A total ban on neurotechnology would be neither practical nor ethical, as it would deny necessary medical treatments to patients who rely on these technologies. Instead, graduated regulation is required: maximum scrutiny for invasive implants, rigorous standards for non-invasive wearables.

At the heart of these debates lies a profound ethical challenge: not merely how to protect highly sensitive personal data, but how to define and preserve the self in an era when memory, emotional states, attention, and even decision-making processes can be digitally recorded, influenced, or commercialized. If the concept of human dignity is to retain its depth and relevance, neural data must not be reduced merely to another economic resource. Rather, it must be acknowledged and safeguarded as an essential element of personal identity and autonomy.

The ethical implications of these developments have prompted experts to formulate the concept of neurorights, like rights to mental privacy, identity preservation, cognitive autonomy, fair access to cognitive enhancements, and protection from algorithmic biases. Europe explicitly recognises physical and mental integrity as fundamental human rights, providing a legal foundation to address the challenges posed by emerging neurotechnologies. Chile has already integrated neurorights directly into its constitution, and several states in the US have updated their privacy laws to explicitly protect neural data. Such developments highlight the growing recognition that safeguarding mental integrity goes beyond conventional privacy concerns.

This becomes even more urgent as we move toward an era in which artificial general intelligence (AGI) may be designed based on human brain architecture itself. The vision, famously advanced by Ray Kurzweil in *The Singularity Is Nearer: When We Merge with AI*, predicts a future where human cognition is closely integrated with AI through brain-computer interfaces, cloud-connected neural systems, and nanotechnology-enhanced cognition. In such a scenario, where minds and machines increasingly converge, the boundaries between biological thought and algorithmic processing begin to dissolve. Protecting mental autonomy is no longer just about defending individual privacy; it becomes a structural safeguard for personhood in a hybrid human-machine future.

As brain activity becomes more modifiable and interoperable with computational systems, our legal and ethical frameworks must evolve accordingly. Neurorights are fast becoming essential protections in a world where our most intimate mental functions could be externalised or enhanced. The future of mental integrity may depend on how clearly we draw the line between augmentation and intrusion and how willing we are to defend that line.

Rethinking AI Ethics

A dystopian future is not inevitable, yet it becomes plausible whenever technical progress outruns moral reflection. AI systems already influence critical decisions: assigning medical treatments, screening job applicants, educating children, making financial investments, and—when paired with neurotechnology—even reading thoughts or modifying human behavior. If these systems remain driven solely by narrow performance metrics—speed, accuracy, profit—we risk overlooking the human values essential to making such decisions ethically acceptable.

Currently, deployers face two key ethical strategies: capability control, which limits what an AI can physically do, and value alignment, which steers what it tries to achieve. Yet both methods have limitations. Restricting capabilities becomes ineffective once AI reaches high levels of autonomy while embedding human values fails if those values are too narrow or biased. To overcome these problems, we need a broader ethical framework, one that teaches AI systems to understand and respect the spirit behind our rules, not just their literal wording. It is critical to understand that no matter how convincingly human-like these systems become, their internal experiences (if they have any) might differ significantly from ours. Recognizing this fundamental difference should reshape our approach: We must not strive simply to replicate human morality in AI, but instead leverage

AI's unique capabilities—speed, scalability, pattern-depth—to complement and enhance human ethical reasoning.

In his influential book *Human Compatible*, Stuart Russell argues that the traditional model of AI, where machines follow fixed, rigid objectives, is inherently risky. He proposes a different approach, suggesting that AI systems should remain fundamentally uncertain about human preferences, continually learning and adapting by observing human behaviors. Rather than fixed commands, AI systems would, thus, aim to align their goals dynamically with evolving human values, making the technology inherently safer and more responsive to genuine human needs. But such a vision also reframes the role of AI: not as a tool that executes commands, but as a system embedded in the moral and cognitive complexity of human life.

This stance forces a wider self-examination. As machines assume roles once reserved for people, we must clarify what must remain distinctly human—empathy, moral judgment, reflective self-critique—and ensure that AI augments rather than eclipses those traits. That, in turn, demands genuine interdisciplinarity. Philosophers, ethicists, historians, and artists must sit alongside engineers, scientists, lawyers, and policymakers so that innovation proceeds in concert with robust ethical insight. Done well, the integration of AI into our society presents an opportunity to improve human cognition, address complex challenges, and promote collective flourishing. Guided by shared values rather than short-term gain, smarter machines can widen the moral circle and advance society instead of merely accelerating it.

Sharing Intelligence

A more nuanced way of thinking about AI involves acknowledging that intelligence, whether biological or artificial, is not inherently threatening. Historically, highly gifted individuals have not generally used their abilities to exploit or oppress others, but rather to advance knowledge and spark innovation for the well-being of society. AI, similarly, need not pose inherent dangers simply due to superior cognitive abilities. The real risk lies in who builds AI and with what data, for what purpose, and under which incentives. When profit-maximization is the sole compass, Big Tech deployment can warp AI away from the public good.

Thus, our goal should not be to constrain or fear AI's growing intelligence, but to cultivate and guide it responsibly, using it as a tool to enhance human cognitive capacities and improve ethical decision-making. AI can be explicitly designed to help us recognize and overcome our biases, inconsistencies, and ethical blind spots. Instead of worrying that AI might inevitably threaten our humanity, we should embrace its potential to elevate human reasoning and amplify moral insight.

Ultimately, the ethical character of AI will mirror the ethical character of the society that creates it. If we thoughtfully approach AI with transparency, in close collaboration—guided by fairness, diversity, safety, and other shared human values rather than market-driven exploitation—smarter machines can become partners in widening the horizon of human dignity and collective flourishing. In such a future, humans and AI will not compete, but cooperate in building a more intelligent, compassionate, and humane world.

About the Author

Gabriela Bar has a PhD in Law and works as an attorney at law and founder of the firm Gabriela Bar Law & AI. After many years of supporting business, she has become an experienced expert in the field of new technology law and the law and ethics of AI. Recognized in Forbes' list of the 25 Best Business Lawyers and the TOP100 Women in AI in Poland, she is a member of Women in AI, the Association for New Technology Law, and the FBE New Technologies Commission. Teaching at several universities, she actively participates in scientific research, conferences, and industry seminars. Gabriela is an author of numerous publications in the field of AI, digital services, and personal data protection. She was a lawyer in EU projects on AI in smart industry: SHOP4CF and MAS4AI. Currently she is an independent AI ethics advisor in the EXTRA-BRAIN project.

Email: gbar@gabriela.bar
Website: https://gabriela.bar/
LinkedIn: https://www.linkedin.com/in/gabrielabar/

CHAPTER 3

AI ISN'T REPLACING YOU—BUT SOMEONE WHO USES IT WILL

By Christopher Bosley, MLS
Machine Learning and AI Consultant
Salt Lake City, Utah

The future of AI is not about replacing humans, it's about augmenting human capabilities.

—Sundar Pichai, CEO of Google

Imagine two mid-career professionals in high-skill jobs. They have worked together for years and have always been highly regarded by their colleagues. Suddenly, one is twice as productive as the other, seemingly out of nowhere. What's the secret? If their employer is looking for who to promote or ways to improve efficiency, which employee will be in the driver's seat? If the secret sauce is AI, which one is leading the company forward and which is being replaced?

The productivity gains inherent in AI technology adoption are so significant that businesses will not pass them up, and adopting these technologies will keep you competitive in business. This is true whether you work in SaaS, retail, eCommerce, gaming, or any other industry. This chapter will walk through the explosive growth of AI solutions in the modern workforce, consider why AI adoption will be non-negotiable for many professionals, and explore a few surprising places where AI is being used. The goal is to arm you with the imagination to augment your work with AI because AI isn't taking your job, but someone comfortable with AI is.

AI Solutions Are Arriving—and Quickly!

The media has been full of AI adoption stories lately, and buzz about the AI revolution is everywhere, signaling that AI solutions are arriving quickly. At JP Morgan, signing and distributing commercial loan agreements is part of the usual course of business. Lawyers review contracts, signatures are applied, and loans are made. Sometimes mistakes are made in the over 12,000 wholesale agreements signed each year, and the 360,000 annual contract review labor hours are a testament to how central contract review is to operations at J.P. Morgan... Until Contract Intelligence (COIN) was introduced. This software reviews contracts in seconds and rarely makes mistakes. A JP Morgan annual report highlighted this impact: "In an initial implementation of this technology, we can extract 150 relevant attributes from 12,000 annual commercial credit agreements in seconds compared with as many as 360,000 hours per year under manual review. This capability has far-reaching implications considering that approximately 80% of loan servicing errors today are due to contract interpretation errors" (JPMorgan Chase & Co., *2016 Annual Report* p. 49 (2017), https://reports.jpmorganchase.com/investor-relations/2016/pdf/2016-annualreport.pdf). The recent wave of AI adoption is full of stories like this. This might sound like a story from the modern AI conversation, but COIN was implemented in 2016.

Artificial intelligence (AI) solutions are in the zeitgeist today, but companies have been developing these technologies for years. They have been, and increasingly are, changing the professional world. In many ways, COIN was a precursor to the current AI boom. This was an early demonstration of the benefits companies could capture with the adoption of AI, and the advent of generative AI solutions has dramatically accelerated this process. This isn't just material from a corporate report. Studies are independently reproducing these results. A 2024 study performed by Onit, Inc. demonstrated that LLM accuracy in contract review is on par with that of trained lawyers and contract professionals and performs at significantly higher speed and dramatically lower cost (Andrea M. Long, Joshua B. Miller, and Joseph M. Johnson, *Better Call GPT, Comparing Large Language Models Against Lawyers*, ARXIV, arXiv:2401.16212 (Jan. 26, 2024), https://arxiv.org/pdf/2401.16212). This is only a single example of a task that requires highly trained, skilled professionals, which is almost trivially performed by an AI solution.

With results like these in high-skill professions, the threat seems obvious: AI is coming to take your job. The truth is somewhat different: Artificial Intelligence isn't going to take your job, but someone fluent in AI tool use is.

Trends in AI Adoption

The use of AI solutions in business settings has grown dramatically. An Accenture study shows that AI adoption has increased, on average, from 30% to over 50% of companies across all industries (Accenture, *The Art of AI Maturity: Advancing from Practice to Performance* p. 11 (2023), https://www.accenture.com/content/dam/system-files/acom/custom-code/ai-maturity/Accenture-Art-of-AI-Maturity-Report-Global-Revised.pdf). Ninety-two percent plan to invest more in the next two years.

Technology is a prominent leader in adoption, but the automotive sector, retail, utilities, media, public services, travel, banking, and capital markets all show a marked increase in AI adoption. Even the smallest advances in penetration are on the order of 10 points. This year, as much as 20% of current technology budgets are allocated to AI efforts (Coherent Solutions, *AI Adoption Across Industries: Trends You Don't Want to Miss in 2025*, COHERENT

SOLUTIONS (May 7, 2025), https://www.coherentsolutions.com/insights/ai-adoption-trends-you-should-not-miss-2025).

The growth in GenAI is even more explosive: 33% to 70% in a mere two years, per a recent McKinsey report (McKinsey & Company, *The State of AI: How Organizations Are Rewriting to Capture Value* ex. 8, MCKINSEY (Mar. 12, 2025), https://www.mckinsey.com/capabilities/quantumblack/our-insights/the-state-of-ai). A Microsoft meta-analysis of GenAI application adoption and task completion showed consistent speed increases of 50% versus individuals not coworking with a GenAI application (Microsoft Research, *AI and Productivity: A Study of Developers in the Real World—First Edition* fig. 1 (2023), https://www.microsoft.com/en-us/research/wp-content/uploads/2023/12/AI-and-Productivity-Report-First-Edition.pdf?msockid=17c621352fa466ed305032e22e396772). A McKinsey report states: "As a result of these reassessments of technology capabilities due to generative AI, the total percentage of hours that could theoretically be automated by integrating technologies that exist today has increased from about 50 percent to 60–70 percent" (McKinsey & Company, *The Economic Potential of Generative AI: The Next Productivity Frontier* (June 2023), https://www.mckinsey.com/capabilities/mckinsey-digital/our-insights/the-economic-potential-of-generative-ai-the-next-productivity-frontier).

In the last eight years, not only has adoption accelerated, but the growth rate in adoption has also accelerated. McKinsey research suggests, "The midpoint scenario at which automation adoption could reach 50 percent of time spent on current work activities has accelerated by a decade." McKinsey & Company, *supra*, at ex. 8. 2017 estimates placed this midpoint in 2053 while their updated model suggests 2045.

GenAI applications have had outsized impacts on collaboration and the application of expertise (McKinsey & Company, *supra*, at ex. 10). These areas were traditionally more resilient to automation. Physical labor, for example, shows little automation potential with the rise of GenAI. Still, decision-making, collaboration, and data management activities are the focus due to their reliance on natural language and interpretation. Moreover, this is also forecasted to impact roles requiring higher degrees of educational attainment (McKinsey & Company, *supra*, at ex. 12.) and higher-wage roles (McKinsey & Company, *supra*, at ex. 13), a further contrast from previous automation revolutions.

AI Adoption Is Non-negotiable

So many companies adopting AI suggest the competitive landscape is shifting quickly. Companies that don't start adopting AI solutions will risk falling behind their competitors in innovation, competing against more efficient processes, and having increasing difficulty attracting top talent that has already augmented their work with AI.

Augment Your Work with AI

In the face of increased adoption, efficiency, and productivity gains, and technology making inroads into work areas traditionally insulated from automation, won't we all just be replaced by AI? Consider the JP Morgan example above. Are some lawyers likely to be replaced by AI? Indeed, companies will always seek to decrease costs where possible. But what about the lawyer at JP Morgan who embraces this technology and increases their productivity significantly. Remember that Microsoft showed consistent speed increases of 50% for individuals working with a GenAI application (Microsoft Research, *supra* at fig. 1). Not only is the lawyer who embraces AI technology working more efficiently, but their ability to co-work with AI solutions is enabling efficiency improvements. The worker who learns to use these tools most effectively will be the worker who is increasing efficiency and not being left behind.

According to a multi-university and Microsoft study, software engineers using AI-augmented workflows demonstrate average productivity increases of 26% (Zheyuan Cui et al., *The Effects of Generative AI on High Skilled Work: Evidence from Three Field Experiments with Software Developers*, SSRN (Sept. 2024), https://

ssrn.com/abstract=4945566). For example, a tool like GitHub Copilot allows an AI to analyze comments, documentation, and code to generate suggestions for relevant code snippets, comments, and documentation. This low-barrier adoption allows the expert (a software engineer) to augment their workflow using GenAI.

According to the research work of Antonios Thanellas and his team: "HUS Helsinki University Hospital, CGI and Planmeca have jointly developed an artificial intelligence solution that assists radiologists and clinical doctors in interpreting brain CT scans and detecting the most common types of non-traumatic brain hemorrhages" (Antonios Thanellas et al., *Development and External Validation of a Deep Learning Algorithm to Identify and Localize Subarachnoid Hemorrhage on CT Scans*, 100 NEUROLOGY (2023), https://doi.org/10.1212/wnl.0000000000201710). This is an exquisite example of AI-augmented high-skill work. In this process, CT head scans are (1) independently analyzed by a trained AI agent, (2) the AI generates information about potential diagnosis and supporting evidence, and (3) radiologists and specialists examine the imaging and AI output to arrive at a diagnosis. This process diagnoses common brain bleeds with nearly 98% accuracy and allows highly trained clinicians to review more cases. Augmenting these professionals with AI agent assistance not only increases efficiency, it also can save lives.

Even everyday office tasks can be augmented with AI. Microsoft Copilot (as well as many other offerings, including Otter, Fireflies, Equal Time, MeetGeek, MeetRecord, Gong, Avoma, and many more) has a dedicated meetings assistant function. It can transcribe discussions, generate notes and summaries, track talking time, note points of agreement or disagreement, and generate action items. The tasks AI can support you in aren't limited to fancy, complex, or expert tasks. Often, they are just mundane items that make your work (or personal) life a bit easier.

You don't have to be an AI expert to use AI. The lawyers reviewing contracts, the radiologists being aided by AI agents, and most developers using AI-assisted coding solutions all have two things in common: They are highly trained experts in their field who aren't necessarily AI experts or engineers. In fact, many AI leaders advocate for human-in-the-loop AI solutions (McKinsey & Company, *"A Human in the Loop Is Critical." McKinsey Leaders on Generative AI at US Media Day*, NEW AT MCKINSEY BLOG (July 3, 2023), https://www.mckinsey.com/about-us/new-at-mckinsey-blog/keep-the-human-in-the-loop). The highest quality and most successful solutions frequently require the nuanced judgment of highly trained subject matter experts. The key isn't understanding the underlying algorithms or writing code to train AI agents. It's about learning the capabilities and limitations of these solutions and how to frame problems in a way that GenAI can support.

Think through the tasks you do every day. Are you developing strategic road maps and translating these into smaller action plans? Are you evaluating a contract for legal issues? Are you examining images to determine how many cars flow through an intersection? All these workflows could be augmented by AI. The best way is to start trying! Certainly, don't feed sensitive or proprietary information into an AI solution, but try out some publicly available solutions. See what ChatGPT, Gemini, Copilot—any of these—can do in your everyday work. You aren't going to break anything by trying these solutions out, and you might start finding a pattern that can make you more efficient. Think about what GenAI solutions are good at: They can interpret language and return results based on prompts. How often do you engage in routine, repetitive tasks vital to your work? Try seeing how these can be automated, and as you develop this workflow, share it with others. Talk to your coworkers, share notes, improve each other's workflows, and bring your technology partners into the conversation. They may be able to construct safe and secure frameworks for regular work use and enshrine this as work technology. Never stop learning. Just by reading this book, you are starting down the path of augmenting yourself with current technology. Don't lose this impulse.

Final Thoughts

AI solutions are coming, and they are coming more and more quickly. Adoption will continue to increase, and more areas of our lives will be impacted. These efficiency gains are significant. AI will be a part of all our professional lives. Adopting technology and augmenting your work with AI makes you a part of the solution, not someone left behind in the wake of the AI revolution. Imagine a future where people can concentrate on what they are good at: creative work, judgments, abstract idea connections, and applying their expertise—and AI can assist them with the mundane tasks computers are better suited to. Instead of thinking about AI as a threat, imagine the possibilities of AI-augmented work using your expertise.

About the Author

Christopher Bosley is an AI and data executive with more than 15 years' experience leading teams and building large-scale, production-ready AI solutions and holder of a Master of Jurisprudence. His expertise spans multiple industries including AAA games, e-commerce, logistics, and SaaS.

LinkedIn: https://www.linkedin.com/in/chris-bosley-aiml/

CHAPTER 4

BETTER TOGETHER: WHY AI'S PROMISE REQUIRES HUMAN CONNECTION

By LingYi Chang
AI Consultant and Speaker, Co-Founder of illumi.one
Singapore

A boat doesn't go forward if each one is rowing their own way.
—Swahili Proverb

In June 2025, a 10-hour ChatGPT outage sent shockwaves worldwide. What followed was more than inconvenience; it was mass withdrawal. News reports described people feeling "as if they had lost a piece of themselves." Many admitted to spending hours each day talking to the AI, using it to process emotions, sort through their thoughts, and, in some cases, simply to feel less alone.

The panic revealed an uncomfortable truth: We weren't just using AI as a tool. We had woven it into the fabric of our thinking, daily rhythms, even our relationships. But this integration came with an unexpected cost, one that most organizations are only beginning to recognize.

A Subtle, Powerful Shift in How We Work

As AI chatbots and digital assistants become as commonplace as email, we find ourselves relying on them in unexpected, deeply human ways. AI helps us take notes, write reports, and answer questions, but it also supports us when we feel unsure, helps us sound more confident, and gives us a boost when we're stuck.

The world has embraced this shift, excited by the promise of better productivity and new creative tools. Yet the outage—and how we reacted to it—forced us to ask whether our embrace of AI has come with an invisible price. Are we quietly losing collaboration, creativity, and healthy friction, forgetting that these fundamentally human qualities have always driven our greatest innovations forward? What will the workplace look like when AI assistance isn't just helpful, but ubiquitous?

The Future of Work?

Let's envision your workplace five years from now. Every new hire receives an "AI crew" on day one. Amy, who wields a research synthesizer, a style-checker, and a competitor-tracker. And John, whose daily rhythm depends on a forecasting analyst, a budget watchdog, and a slide-deck creator.

On the surface, the arrangement appears idyllic: higher productivity, fewer errors, happier employees. But there's this underlying emptiness, as if the office has lost its heartbeat. Because nobody talks to each other anymore.

Amy asks her AI about competitors instead of talking to Tom, who just wins over them in a deal and knows their weak spots. When John hits a problem, he brainstorms with AI instead of pulling his team together to work it out. Somehow, all those pantry conversations that used to happen throughout the day just... disappeared.

We thought COVID-19 created distance between people. But this might be worse. Back then, we felt cut off from each other and made a real effort to reconnect. Now people are choosing isolation because it seems easier. Why bother pinging a coworker or risk an awkward interruption when your AI is always available, always fast, never tired, never grumpy?

The data paint a disturbing picture. According to BCG's latest research, innovation readiness among companies dropped from 20% to just 3% between 2022 and 2024, precisely as ChatGPT launched and AI adoption exploded. We expected AI to fuel innovation. Instead, we're seeing the opposite.

The Twin Traps of AI Dependency

What's really happening here? Underneath lies a tension between two seductive, ultimately self-defeating traps.

The Seduction of Going It Alone

Let's be honest, what's more alluring than the idea of becoming an AI-powered superhuman? LinkedIn influencers and tech marketers sell this dream relentlessly. Who needs messy meetings or team friction when you can delegate to AI that never needs sleep?

But this fantasy ignores how true intelligence and productivity emerge. A recent 2025 METR study found experienced software engineers using AI coding tools were actually 19% slower than those working without AI, despite believing they were 20% more productive. This isn't an isolated finding; it reflects a deeper pattern: AI tools promise efficiency but often deliver complexity, cognitive overhead, and lower quality.

Why do smarter tools so often leave us spinning our wheels? Research suggests that when individuals chase efficiency above all else, they often sacrifice depth and quality. The more projects a single person juggles, the less likely any one of them will break new ground, a principle that extends beyond individual work to how we structure innovation itself.

Here's what the AI superhuman fantasy misses: Breakthrough innovation has never been a solo sport. Just like Steve Jobs needed Jony Ive's design eye and Tim Cook's operational discipline. Marie Curie leaned on Pierre and a laboratory of technicians whose names rarely grace the textbooks. The myth of the self-made innovator is comforting, but it is just a myth. The greatest achievements of the past century—from scientific breakthroughs to social revolutions—have always been powered by the collective genius and friction of diverse teams. Lone wolves may get our attention, but it's the pack that wins the hunt.

When Help Becomes a Crutch

If the AI-powered isolation is the first trap, the second is cognitive atrophy. A recent MIT study highlighted this concerning trend: The more we rely on AI for creative work, the less we exercise our own problem-solving abilities. It's like having a personal trainer who does your push-ups for you: It's convenient, but ultimately weakening.

But the implications go deeper than individual skill loss. Robin Dunbar, a renowned Oxford professor, discovered through his social brain hypothesis research that human intelligence itself evolved through collaboration and social interaction. Our brains didn't develop their capacity in isolation; they grew powerful by managing complex relationships, navigating group dynamics, and solving problems collectively. When we retreat into AI bubbles instead of engaging with fellow humans, we're not just missing out on better solutions—we're fighting against millions of years of evolutionary wiring that made us intelligent in the first place.

This erosion is already reshaping workplaces in subtle but profound ways. Given this unstoppable trend, how do we stay competitive without falling into the same trap? The answer isn't to abandon AI. That ship has sailed. Instead, we need to rethink how we implement it. The organizations that will survive and thrive are those that use AI to make their people more powerful, not more replaceable. But what does this actually look like in practice?

The 70% Equation—Why Culture Trumps Code

McKinsey dropped a bombshell: 92% of companies invest in AI, but only 1% achieve mature results. Why? The reason isn't technical, in fact, technology is the smallest determinant of AI success. BCG's research reveals that algorithms matter for just 10% of success, tools for 20%, but people, processes, and culture drive 70%. In my experience, culture is what really makes or breaks AI efforts.

Unlike typical IT projects, AI requires business teams to truly co-own the journey to shape how technology fits into everyday work. Consider this: How many customer chatbots can you name that are truly amazing and reliably solve your problems? Very few. Why? Because real success comes from teams constantly refining the experience based on real-world feedback. One of my previous successful projects required a dedicated in-house team of more than 30 people to continuously improve their dialogue flow based on user behavior. This is often a level of investment most companies aren't prepared for.

Of course, with the advancement of large language models, the effort might be reduced nowadays, but the mindset remains. AI projects require monitoring and many trials and errors because no one model fits all. You'll only know how good and how well the model fits after you test it against your data and scenario.

I've watched organizations that spend months perfecting tech designs and gathering requirements from user teams still encounter very low adoption rates. How to avoid these pitfalls? Instead of building first and asking for adoption later, let's get the whole team experimenting with rough, imperfect versions from week one. When people help test the AI, when they see it fail and improve, when they understand its limitations and strengths through hands-on experience, adoption becomes natural. It's not about the sophistication of your final solution, but whether your team has confidence to use the solution, which comes from learning together from day one.

The Beautiful Imperfection of Collaboration

This leads to a counterintuitive insight: Imperfection invites collaboration. When something's rough around the edges, people feel like they can jump in and help. They see gaps to fill, improvements to suggest, connections to make.

By contrast, AI-generated work looks deceptively perfect. Although in many cases AI work isn't actually perfect, it just *looks* perfect. But when everybody brings their AI output to a meeting, the clean look covers up shallow thinking, generic insights, and logical gaps that would be obvious in a rough draft only if we took a deeper look. The professional appearance successfully creates a false sense of completion.

This surface-level sophistication is particularly dangerous in teams. When everyone brings AI output to a meeting, nobody wants to mess with work that looks professional. So, they hold back their natural instincts to question and probe deeper.

And now your *teams* fall into the same cognitive atrophy trap. They gradually lose their ability to distinguish between surface polish and substantive quality. They become consumers of AI output rather than active thinkers, nodding along to content that sounds smart but lacks depth. We forget that people's raw, honest feedback is usually where the best breakthrough ideas are born.

To counter this, teams need to deliberately cultivate skepticism toward polished AI work. This might mean establishing norms like, "If it came from AI, we question it harder, not less," and accepting that it won't be perfect right away. To do this, teams need to work together even more closely to try things and learn from failures, and discover what works for them. Because innovation usually starts as rough, half-baked ideas that invite input and improvements.

From Prompts to Context: The Team Intelligence Revolution

Most AI implementations today are fundamentally individual-focused. You open a chat window, describe a task, get a generic response, and spend the next hour editing and refining it. This is because the system knows nothing about yesterday's decisions, your brand voice, or last quarter's lessons. The "back and forth" waste shows up in hard numbers: Atlassian's 2025 State of DevEx survey found that 68% of developers now save at least 10 hours a week with AI tools, but 90 percent still lose six hours or more (and half lose over 10) chasing missing context, waiting for answers, or untangling duplicate work. In other words, the gains from faster code are being canceled by the drag of scattered knowledge.

The root problem isn't the quality of the prompts; it's the poverty of the shared memory those prompts can draw from. Prompt engineering is like texting a "professional stranger" for help: You have to explain everything because they have no idea about who you are or what you're trying to do. This is why context engineering is starting to attract attention. Done well, it teaches the AI to participate in the team's ongoing dialogue, tapping into collective experience, strategy, and nuance before any question is asked. When this "professional stranger" understands your goals, constraints, and tone, they become a true collaborator. Their first draft would land far closer to final, and the team can refine instead of rebuild.

That's why I believe we need to reframe AI not as a solo craft, but as a collaborative team process. That principle sits at the heart of illumi, the visual collaboration platform my co-founder and I built for teams working with AI. Instead of isolated chats, every note, prompt, and AI output lives on a canvas everyone can see and refine. As soon as one person learns something, the whole space learns with them, and AI's next answer builds on what everyone has learned together. Teams using illumi report shorter learning loops, higher-quality drafts, and—maybe most important—stronger interpersonal bonds because the thinking process itself is now visible rather than hidden in private threads.

Rewiring the Workplace: Building the Collaborative AI Culture

Making this shift is as much cultural as it is technical. I see the following common characteristics from teams that embrace AI most effectively:

- *They start with real problems, not AI solutions.*

Before jumping into implementation, teams deeply understand what they're actually trying to solve: workflow bottlenecks, knowledge gaps, or communication issues.

- *They share rough workflows and early models, not polished demos.*

When Amy shares her half-baked AI prompt for analyzing customer feedback, Tom can suggest improvements and adapt it for his market research work.

- *They lower barriers, so anyone can contribute improvements, not just engineers.*

Often the best insights come from unexpected sources, like the sales rep who notices AI consistently missing cultural nuances in customer communications.

- *They reward learning cycles and visible evolution, not just ticking checkboxes.*

Teams that discover 10 ways an AI approach doesn't work have saved everyone else from those same dead ends.

Trust, Relationships, and Human Values

As AI's role grows, so does the need for trust and transparency. Emotional safety and shared identity grow more important, not less, in an AI-powered workplace. When machines can replicate many human capabilities, what becomes precious is the uniquely human ability to create meaning together, to build trust, and to navigate ambiguity as a team.

Atlassian stated it: Successful teams don't just use new tools; they innovate how they work. The teams that will thrive in the AI era treat every project as a test of how to work better together. This collaborative approach matters more than ever because today's workplace is increasingly fluid. People no longer commit decades to a single company. Remote and hybrid work means team members might be scattered across continents. The old ways of bonding—working in the same office for years—are going away. Now, being able to quickly connect and work well with new people is essential.

The most effective teams in the AI era will be those that master collaboration, groups that can gel quickly, experiment, and adapt their working methods as the technology evolves. When these dynamic teams collaborate on AI implementation, they create something no individual could achieve alone: a collective intelligence that's greater than the sum of its parts.

The Call to Action: First Steps Toward Collaborative AI

The Swahili proverb that opens this chapter is more than poetic wisdom; it's practical advice. AI can give us stronger oars, but it can't set the rhythm of the crew. That rhythm comes from people who choose to work together. Here's the good news: You don't need a perfect plan to get started. Here are some practical, down-to-earth ways to get your team experimenting with AI:

- *Start ridiculously small.*

Pick one repetitive task everyone hates, and try AI on it together. Share the messy first attempts and laugh at the weird outputs.

- *Make it a 15-minute Friday thing.*

End your week by having someone demo what they tried with AI—good, bad, or hilariously wrong. No pressure, just "Here's what happened when I asked ChatGPT to help with this."

- *Pair up skeptics with enthusiasts.*

Don't force it, but when someone's curious about AI, match them with a colleague who's already tinkering. Learning together removes the intimidation factor.

- *Create a simple sharing channel.*

People will naturally start sharing screenshots of both their successes and the times AI completely missed the mark. The fails are often more valuable than the wins. These small moments of openness build into something remarkable: a team that learns faster and has more fun doing it.

Tools will get better, models will grow more powerful, but the practice of learning together will always enable teams to go further than individuals working alone. The most successful teams will move forward together, each step powered by AI but guided by something wonderfully human: our ability to connect, adapt, and create something amazing as one.

References

1. Innovation Systems need a reboot https://www.bcg.com/publications/2024/innovation-systems-need-a-reboot

2. Your brain on ChatGPT https://www.media.mit.edu/publications/your-brain-on-chatgpt/

3. The Social Brain Hypothesis and Its Implications for Social Evolution https://www.researchgate.net/publication/26338803_The_Social_Brain_Hypothesis_and_Its_Implications_for_Social_Evolution

4. https://metr.org/blog/2025-07-10-early-2025-ai-experienced-os-dev-study/ https://www.microsoft.com/en-us/worklab/work-trend-index/ai-at-work-is-here-now-comes-the-hard-part

5. https://www.mckinsey.com/capabilities/mckinsey-digital/

our-insights/superagency-in-the-workplace-empowering-people-to-unlock-ais-full-potential-at-work

About the Author

LingYi Chang (Ling) is a technologist and entrepreneur with 15 years of experience, including 8 years leading AI and machine learning initiatives at Microsoft and AWS across Asia-Pacific. She excels at translating complex technology into practical business solutions, helping companies solve problems and drive growth.

Ling serves as a trusted advisor to leading consulting firms and financial institutions. As a recognized industry voice, she regularly speaks at events for Nvidia, INSEAD, UBS, and more, advocating for transparent, collaborative AI approaches that accelerate organizational learning.

As co-founder and Chief Product Officer of illumi, Ling identified how siloed AI implementations limit collaborative potential. Her company developed a visual collaboration platform enabling teams to collectively experiment, learn, and build with AI in shared digital environments.

Ling holds an MSc in Management from Imperial Business School, England. Born in Taiwan and based in Singapore, she promotes balance between innovation and wellness as a certified yoga teacher.

Email: ling@illumi.one

Website: https://illumi.one

LinkedIn: https://www.linkedin.com/in/lingyichang/

MARKETPLACE FOR INTELLIGENCE: AN ESSENTIAL DIFFERENTIATOR FOR INTELLIGENT ENTERPRISES AND INDIVIDUALS

By Naser Chowdhury, PhD
Applied AI and ML Practitioner; Technology Architect
Phoenix, Arizona

Within the next five years, any economically valuable job humans can do, AI will be able to do 80% of it...80% of all jobs can be done by an AI.

–Vinod Khosla (Tech Entrepreneur and Investor)

The skill gaps between human workers and AI agents are shrinking every day, to the point that experts predict almost 80% job loss among human workers by 2030, while only a low single digit of companies currently claim to have reached an AI implementation maturity level. While the trend of companies improving productivity and reducing costs through an AI-agentic workforce may not change in the foreseeable future, there are few essential steps that governments, individuals, and companies can take to ensure they are all winners in the long run. The first step is to realize and recognize that these AI-induced changes are fundamental, and time is not on their side. The second step is to ensure they know how to utilize AI and sensor-powered machine intelligence, as well as robotics, to their advantage, making themselves much more valuable than a machine or a human alone. The third step is to start recognizing "intelligence" as a currency and to create marketplaces where human, machine, and artificial intelligence are traded, just like physical or digital goods, for value. The only way to stay ahead in the AI, smart machines, and robotics era is not to ignore their presence, but instead to embrace them, integrate them into every step of human life for better decision-making, and outcompete the competition.

Introduction

Imagine you are the CEO of a Fortune 500 company, and you wake up to the news that Silicon Valley investor Vinod Khosla predicts AI will replace 80% of jobs by 2030, taking much of the Fortune 500 with it. This sobering news might not entirely shock you as much as it does the mid-career professional who has a growing family to care for and is not necessarily in a position to start a new academic or professional training journey.

Now imagine the impact of a newly graduated college grad who is about to join the workforce or an owner of one of the millions of small and medium-sized businesses that are barely getting by with their bare minimum sales. They will immediately start thinking that without jobs, their customers would be unable to afford their products and services. The most critical number likely sticks to their mind, 2030 which is in five short years or less! They will start to think about what they, their families, and friends need to do to stay relevant to their societies and themselves, before it's too late.

Now, the next news from the same morning is encouraging, as NVIDIA's CEO Jensen Huang said, "AI is the great equalizer. It's helping me become better and better informed in a lot of different fields." It must be true, and NVIDIA's current market value of over $4 trillion is a testament to that. He continued his recent discussion with MSNBC news, stating, "My job has already changed. The work that I do has changed, but I'm still doing my job." True, but how he is doing it repeatedly and successfully becomes more of a viral search question when more news shows up in the news sites and inboxes, like Microsoft laid off 6,000-plus workers, Salesforce laid off 1,000-plus workers, Dell almost 12,000 workers, and HP 2,000-plus workers, with the headlines: "The AI Layoffs Began"!

The purpose of this article is not to bring the inevitable truth of AI's impact on the job market, societies, or individuals, but to get some new innovative thinking in the face of such a life-altering situation concerning what individuals, small business owners, or government can do. By doing so, instead of standing still in the face of these income-shrinking, lifestyle-shattering, technological advancements, we can prepare in a way that is equally strong to withstand any current and future threats on human dignity.

Five Years: Targeted Timeline for Major AI Transformation for Enterprises

In the AI era where everything is directly or indirectly operated by some level of intelligent systems, machines, or algorithms, it is not the question of whether or not enterprises will need to be completed or rearchitected to match the demand of the AI era, but when. When will the transformation of enterprises need to be completed? Enterprises whose growth started incrementally with cloud, Agile, DevOps, and digital now need to become part of a successful AI transformation powered by LLMs, intelligent agents, chatbots, and intelligent assistants.

At a glance, it might appear that AI-powered agents, assistants, and their autonomous course-correcting skills will overtake corporate roles immediately. Human roles seem about to become redundant, but the reality is that today's enterprises are multidisciplinary, complex systems that will require time and human expertise before they become fully autonomous, especially in areas such as mission-critical business and engineering operations.

In addition, a 2025 study by McKinsey found that the pace of corporate AI adoption is not uniformly advanced and not equally prioritized; therefore, the lead time of five years before significant job-impacting changes is not impractical, but somewhat relevant and valuable information. This same McKinsey study found that the most AI-impacted corporate divisions, with annual revenue of $500 million, include risk, compliance, data governance for AI, AI strategy, roadmap for AI-enhanced or AI-focused products, tech talent, and adoption of AI solutions.

Anatomy of an Intelligent Enterprise

The hypothetical building blocks of intelligent enterprises comprise four major areas and four sub-areas, all of which play crucial roles in the enterprises. Each C-level executive plays a critical role, and the CEO oversees everything.

- *Chief Operating Officer (COO)*: Mostly responsible for orchestration, collaborations, operations, service design, business strategy, and the execution of technology strategies
- *Chief Technology Officer (CTO)*: Responsible for the enterprise's artificial intelligence, machine learning, agentic AI, generative AI, robotics, software development, robotic process automation, AI capital management, and R&D leadership
- *Chief Information Officer (CIO)*: Responsible for IT infrastructure, applications, agents, data, hosting, platform, benchmarking, and performance

- *Chief People Officer (CPO):* Manages human capital, policy, governance, compliance, and strategic vision for human capital utilization

In addition to above critical roles and responsibilities in modern intelligent enterprise, the CEO ensures the organization executes stellar strategic vision, overall cloud, big data, ETL, and API asset management, empowers both AI and IT initiatives, ensures workflow automations are flawlessly designed and executed, and sees that a hybrid intelligent system is designed and operated in which human intelligence is utilized at the highest level of efficiency with the integration of robotic, artificial, and machine intelligence.

Key Technology Enablers of Intelligent Enterprises

In recent times, the most impactful technology in enterprises has been LLMs (large language models) and the applications they enable. Trained with billions of parameters and knowledge from the entire web, these large language models made many humans' acquired knowledge and skills redundant. Gone are the days when researchers had to scour the web and sift through research articles daily to gather the most relevant information.

Today, answers to many research questions are at the fingertips of almost any individual with advanced systems, such as retrieval augmented generation (RAG), and prompting capabilities, thanks to the availability of vector databases and semantic searches. Agentic AI and multi-agent systems are the new standard for many corporations, and an agentic workforce that works with minimal to no human intervention is the new desired technological capability of many C-suite executives. Cloud, big data technologies, and predictive analytics started the AI revolution, and LLMs with reasoning capabilities are making those complex implementations possible. Vibe coding, also known as prompt-based coding, makes prototyping easily accessible and time-saving solutions for many.

Meanwhile, tools that can create production-grade software, agents, small industry-specific models, and infrastructure are building the new competitive edge among enterprises. Relevant and high-quality data remain in high demand to produce high-quality AI-generated results. At the same time, security concerns and hallucinations by LLM models continue to keep CTOs cautious and grounded when they seek to leverage the full potential of AI.

Gaps Between Intelligent Enterprises and General Enterprises

Today, many small and medium-sized businesses (SMBs) and some enterprises are not ready for the AI revolution. Current operational, technological, and talent stacks, as well as software and hardware, require knowledge and skills to integrate new AI and IT solutions into their technology and business stacks. The reason why these gaps exist is multi-fold. The engineers and skilled workers who make these AI technology implementations possible for enterprises are in short supply. There are not enough qualified workers available in academia or industry who are affordable for enterprises across the board.

In addition, because the technology is new and innovative, and its reliance on traditional IT solutions is complex and interconnected, AI transformation is not always as straightforward as it appears to be. There are too many unproven claims from service providers and not enough auditors or testers to validate those claims. The burden of proof makes AI transformation even more difficult for organizations. In addition, new regulatory and compliance-related responsibilities are making enterprises more cautious when road-mapping AI solutions for their organizations. Therefore, the gaps between AI skill workers and enterprises that need them are fundamental. At the same time, human talents are getting laid off due to AI implementation and automation, which are also a reality.

Unfortunately, due to the rapid implementation of AI technology, the advancement of their proprietary technology has made knowledge transfer of AI implementation skills one of the most significant barriers at the current time, as academic training without practical industry skills is difficult to fill. Almost every company has access to Microsoft Office, which does not necessarily mean that every company is equally successful. Similarly, nearly every company can have similar general-purpose AI solutions and agents. Yet, results might still vary due to differences in training, implementation experience, and the skills of the workforce. Therefore, there is a massive opportunity for AI technology service providers, as well as individuals with the right skills, knowledge, and experience, to capitalize on these skills and experience gaps and prepare themselves for the enterprises of the future.

Introducing the Concept of Marketplace for Intelligence

AI technologies and innovations are transforming businesses and industry landscapes every day, worldwide. The technology changes and possibilities are massive, with implications for the industry landscape not only in one industry but across categories of multiple industries and domains, and their cascading effects are equally impactful. On one hand, not every company has an AI R&D department due to resource or talent shortage. On the other hand, there exists hardly any company that does not want to reduce their labor cost or want to take advantage of revolutionary AI technologies that pose existential threat to companies with the potential to make companies irrelevant regardless of their industry setting. Therefore, it is safe to say that for the foreseeable future, the real differentiators for organizations of all sizes, regardless of industry, will be "intelligence."

There arises a need for a marketplace where "intelligence" is bought and sold like physical and digital goods. Solutions that a simple web search or LLM-based chatbot are unable to provide, information and insights from hours at LinkedIn or VC memo emails are unable to connect and are personalized for specific company needs. Vision that hundreds of random newsletters are unable to maximize and meet. These proposed two-sided intelligence marketplaces are targeted at a particular industry. There's a need for industry-specific news and innovations, research and development, and information about ground-breaking technologies that have the potential to change an industry or group of companies. There's a need for direct connection with innovative companies and solution specialists that are waiting to impact organizations and businesses. There's a need for companies to develop agents and bots for industry-specific problems. Consumers in the marketplace will collaborate with such a platform's or marketplace's intelligent assistants to discover potential matches for their technological and business needs and, in the process, outcompete their competition. In this solution approach, companies do not need to hire large AI R&D teams or analyst teams because the intelligence marketplace works for their consumers and providers 24/7. It's like getting *Intelligence as a Service* ®.

Speed Is the Only KPI Intelligent Enterprises Will Need Going Forward

In today's AI-powered world, a one- or two-week sprint of product development via the Agile software development process is outdated. Gone are the luxuries of testing products for two to three weeks before they go into production. Today, product owners can code their way to almost market-ready products. AI engineers ensure that products are scalable, available, and accessible globally, with proper security, eliminating the need for multi-day or hourly launch events.

Unfortunately, these skills, tool availability, or expertise showcases are not available in today's traditional search engines, trade shows, or week-long corporate request for proposal (RFP) or request for information (RFI) processes. By the time traditional companies gather their requirements and get ready to distribute their RFI or RFP, intelligent enterprises are already prepared to deliver their second or third iteration of similar products to new consumer bases with speed and accuracy not available or possible before. Therefore, a new kind of supplier showcase is necessary to meet the demand of the latest AI era, where speed is the only KPI in the future, and the opposite of speed is nothing but assurance that enterprises are not ready for the next stage of innovation and product visions.

A Vast Majority of Human Intelligence and Potentials Are Untapped

In the face of existential threats, including massive unemployment, loss of human income potential, and lifestyle changes, billions of people on the planet are engaged, utilizing their valuable intelligence through gaming, entertainment, and social media engagement with little to no economic and financial incentives. They are witnessing the reality of how advanced AI technologies are replicating their jobs and careers, and how thousands of jobs are getting cut every day. However, due to a lack of awareness or realization, the majority of the population believes they are and will be immune to AI-impacted lifestyle changes in their lives. The lack of preparedness obscures the fact that the time has come for individuals, societies, and businesses to prepare for the eventuality of AI's impacts and find creative ways to utilize their valuable asset, intelligence.

This reality opens potential for innovative companies and entrepreneurs to invent solutions on how to tap into a vast pool of unused human intelligence for the broader good of society. In the process, they can train human intelligence with AI or machine intelligence, such that a powerful hybrid intelligence system emerges that can positively impact human lives in the decades to come. Education systems, vocational skills, and self-paced training need to prepare humans from all societies to elevate their intelligence level to the next stage, so that low-skilled, low-cost robots, AI agents, or systems can't threaten human life and livelihood. The marketplace for intelligence can create that awareness and prepare human societies for prosperity and growth.

Conclusion

The AI-powered future is not human-free or without human participation. Not at least for the near term, as long as humans can see their potential and prepare themselves for the challenges by improving the skills necessary to remain relevant in a future AI- or machine-powered world. Reid Hoffman and Greg Beato's book,

Superagency: What Could Possibly Go Right with Our AI, suggests that when workers supercharge their creativity and productivity, they essentially create a superagency. Unfortunately, highly impactful intelligence utilization tools and solutions are not always found on regular office-approved software lists or among vendors that provide custom solutions for enterprises or institutions.

It is the responsibility of each individual to increase their AI literacy, for enterprises to discover actual differentiating products, solutions, and talent in the form of intelligence, and for governments to ensure their citizens have adequate training to meet the demand of new technology era, because the impact may easily be either additional tax revenue or distribution of unemployment benefits. The proposal or creation of an intelligence marketplace where human, artificial, and machine intelligence are showcased and traded in the form of agents, assistants, or human subject matter experts in a time-bound price structure is not only the demand of time, but necessary elements for survival in the advanced technical empowered by AI, smart machines, and robots.

The future of individuals, corporations, and society will depend on how effectively intelligence is used, traded, and utilized when human dignity faces existential threats or is overshadowed by profitability and underutilized human intelligence. Intelligence can act not only as the most significant human asset but also as our savior.

About the Author

Dr. Naser Chowdhury, PhD, is a 20-year industry veteran with experience in multiple industries. He earned his Bachelor of Science in Aeronautical and Astronautical Engineering from the Ohio State University, his MBA from Marylhurst University, and his Doctorate in Information Technology with a focus on cloud, big data, and AI. Dr. Chowdhury's dissertation, "Factors Influencing the Adoption of Cloud Computing Driven by Big Data Technology: A Quantitative Study," has been recognized by the US Department of Education.

Dr. Chowdhury is also a Presidential Award Winner for Excellent Academic Performance at Capella University on multiple occasions. Dr. Chowdhury was the first to introduce the trademarked term "Intelligence as a Service®." Dr. Chowdhury earned the MIT Certification in Digital Platforms: Designing Two-Sided Markets, from APIs to Feature Roadmaps. Dr. Chowdhury has experience working with tech giants such as Apple, Verizon, Toyota, Ford, American Express, Enbridge, SCE, UTC Aerospace Systems, and numerous other industry leaders. Currently, Dr. Chowdhury has dedicated his research and mission to elevating human life by integrating human intelligence with AI, machine intelligence, and creating hybrid intelligence that can withstand any lifestyle-changing threats to human lives.

Email: naserchowdhury1@gmail.com.com, for AI consulting and hybrid intelligence-enhancing projects

LinkedIn: https://www.linkedin.com/in/naserchowdhury/

CHAPTER 6

REIMAGINING HEALTHCARE: A TRANSFORMATIONAL CALL TO ACTION

By Will Conaway
AI Advisor, Healthcare Strategist, Educator
Charlotte, North Carolina

The reason why it is so difficult for existing firms to capitalize on disruptive innovations is that their processes and their business model that make them good at the existing business actually make them bad at competing for the disruption.

—Clayton M. Christensen

The Challenge of Disruptive Innovation in Healthcare

The quote by Clayton M. Christensen encapsulates a pervasive dilemma within the healthcare landscape: the paradox that firms adept at existing processes often find themselves ill-equipped to engage with innovations that disrupt those very processes. This friction is particularly pronounced in healthcare, where a myriad of regulatory frameworks, entrenched business models, and an ethos of caution inhibit evolution. To thrive in this unprecedented era, it is paramount to identify and confront the multifaceted obstacles that obstruct innovation while leveraging emerging technologies like AI for transformative outcomes.

Defining the Problem and Confronting the Crisis: An Urgent Wake-Up Call

The healthcare industry in the US is teetering on the brink. Between skyrocketing costs, administrative bottlenecks, chronic disease overload, and growing inequities, our current systems are cracking under the pressure. Patients find themselves ensnared in endless paperwork and fragmented care. Providers grapple with burnout while insurers struggle with rising claims. In this reality, traditional fixes simply don't cut it anymore.

Nearly 60% of American adults live with at least one chronic condition, such as diabetes, heart disease, obesity, or depression, the very conditions that consume nearly 90% of the nation's healthcare filing each year. This burden strains hospitals, insurers, and taxpayers alike. What's more alarming is the widening gap in access among marginalized communities. The promise of equitable care is fading fast.

Yet, amid the chaos, a beacon of hope shines brightly: artificial intelligence. The tools to revolutionize healthcare aren't futuristic; they're already here. They're ready to reshape the patient and provider experience forever.

Statistics That Matter—The Power of Data: Alarming Stats and the Promise of AI

Numbers don't lie, and these are startling:

- $5 trillion: By the end of 2025, US healthcare spending is expected to approach $5 trillion; that's over 18% of GDP.

- 50%: Americans living with at least one chronic disease during their lifetime.

- 90%: The aforementioned chronic conditions, such as heart disease, diabetes, and obesity, account for approximately 90% of healthcare spending.

- 95% of Medicare users have at least one chronic condition, and nearly 50% have four or more.

- McKinsey projects an incredible, up to $150 billion in annual savings by 2026, all due to AI-driven efficiencies.

- AI-powered patient monitoring can reduce readmissions by between 30% to 50%.

- Automation tools may slash administrative tasks by 25% to 40%, freeing up clinician time.

These aren't theoretical numbers; they're outcomes emerging from early adopters and pilot programs. AI holds the key to transitioning from reactive healthcare to proactive wellness, from inefficient systems to streamlined processes, and from fragmented access to equitable care. As AI-powered analytics help identify at-risk and underserved populations, they open doors to targeted interventions and equity.

Historical Context: Why Past Efforts Haven't Captured Lightning

Unlike retail, finance, or logistics, the healthcare industry has lagged in adopting innovation. That hesitancy isn't accidental; it's rooted in structure, safety priorities, and systemic friction. The sector has long been characterized by its cautious approach to adopting new technologies compared to other industries such as finance, retail, or manufacturing. This reluctance stems from a complex interplay of historical, cultural, regulatory, systemic factors and past failures. In this analysis, we will explore these contributing factors and their implications on the healthcare landscape.

Regulatory Deceleration

The FDA, CMS, and HIPAA, all crucial for safeguarding patient welfare, have built-in layers that slow innovation. Every new tool must clear clinical, financial, and security hurdles. While necessary, these processes often feel like speed bumps impeding progress.

Systematic Fragmentation

Consider the stakeholders: hospitals, primary care groups, specialty clinics, payers, pharmaceutical companies, laboratories, pharmacies, and, of course, patients. Each life in its own silo with distinct workflows. One-size-fits-all solutions fail to fit anyone, creating reluctance to risk disrupted workflows.

Culture of Conservative Care

Because human lives are at stake, healthcare professionals stay cautious, and rightly so. But that risk-averse mindset often slows change, even when new tools promise better outcomes with fewer mistakes. Healthcare's go-to instinct is "show me proven data first," which, ironically, slows the pace of proof.

Protectionism of Privacy

Medical data is gold for threat actors. Ransomware, breaches, and leaks drive organizations to lock down. Implementing new tech becomes a negotiation with a wall of compliance and security protocols, adding costs, time, and friction. Bottom line: The past was defined by safety first, but safety can't be the culprit that destroys our future health system.

Past Technology Failures

Past issues with electronic health records (EHR), big data, and other technology initiatives have created substantial hesitance in the healthcare sector regarding the adoption of AI. Many organizations experienced significant challenges during the implementation of EHR systems, including interoperability issues, data accuracy concerns, and user dissatisfaction, which led to frustration among staff and disrupted workflows. These experiences have instilled a cautious mindset, as providers are now wary of adopting new technologies without clear evidence of their benefits and ease of integration.

Additionally, the complexity and variability of big data systems have highlighted the difficulties in managing and analyzing vast amounts of information, making stakeholders wary of committing to similar investments in AI. To complicate matters, most technology projects in healthcare are wrongly seen as "IT projects" and not healthcare business projects; this results in poor executive support. Consequently, these past experiences may lead to a more conservative approach towards early adoption of AI solutions, as organizations prioritize stability and proven effectiveness over the allure of cutting-edge innovations.

The Breakthrough and Imperative for Improvement: What AI Can Do Now

AI in healthcare isn't a concept; it's happening today. Here are just a few examples highlighting real-world impact.

- *Diagnostic Insight and Early Detection*

AI algorithms analyzing X-rays, MRIs, or CT scans can detect subtle signs faster than the human eye. Mobility-impairing fractures, early-stage cancer, and even tiny strokes can be rapidly identified by AI, improving diagnoses and shortening treatment timelines.

- *Operational Efficiency*

AI automates scheduling, prior authorizations, billing, and claims. It helps detect fraud while letting clinicians dedicate more time to patients and less to typing.

- *Chronic Care Management*

Imagine personalized care plans that adapt in real time based on real-world data: heart rate, glucose trends, and activity levels. AI-driven interventions can prevent crises and hospitalizations before they happen.

- *Equity and Outreach*

Advanced analytics spotlight at-risk communities. AI-powered telehealth reaches remote or underserved areas. Mobile clinics guided by AI create novel access points where hospitals don't exist.

- *Administrative Automation*

Machine learning can transform claims reviews, fraud detection, and denials, freeing clinical and clerical staff while enabling financial sustainability, potentially leading to lower premiums or rebates for patients.

Embrace or Dispel: Dramatic Divergence

Here are several areas where the path chosen will dictate future relevance for hospitals.

The Hospital: Diagnostic accuracy will sharpen. Or stagnate.

- *With AI*: Hospitals that deploy AI-powered diagnostic tools will see a dramatic boost in accuracy and speed. Machine learning algorithms already outperform human radiologists in detecting certain cancers and can process thousands of medical images in minutes. AI can analyze symptoms, histories, and genetic profiles to help clinicians arrive at faster, more accurate diagnoses, reducing misdiagnosis rates and improving outcomes.

- *Without AI*: Hospitals that continue to rely solely on traditional diagnostic methods will lag in accuracy. As competitors leverage AI to offer more precise and earlier diagnoses, these hospitals will lose patient trust and referrals. Diagnostic delays or errors will become more glaring in comparison, potentially leading to legal liabilities and reputational damage.

Operational efficiency will skyrocket. Or collapse.

- *With AI*: Administrative burdens are a primary cause of burnout and inefficiency in hospitals. AI can automate scheduling, billing, resource allocation, and even documentation through natural language processing. Hospitals that implement AI will streamline operations, reduce overhead, and optimize staff deployment, all while increasing patient throughput without sacrificing quality of care.

- *Without AI*: Hospitals resisting AI will find themselves drowning in paperwork, manual scheduling errors, and inefficient use of staff and resources. As labor shortages grow and costs continue to rise, these hospitals will struggle to maintain profitability. Operational bottlenecks will drive down patient satisfaction and increase employee burnout, feeding a dangerous cycle of inefficiency and attrition.

Personalized medicine will flourish. Or flounder.

- *With AI*: Hospitals that harness AI to analyze vast datasets, ranging from genomics to electronic health records, will lead the charge in developing personalized treatment plans. AI can identify which therapies are most effective for individual patients, predict complications, and recommend tailored interventions. This will transform care from reactive to predictive, boosting both outcomes and patient satisfaction.

- *Without AI*: Hospitals that stick to generic treatment pathways will fall behind as patients begin to demand tailored care based on their unique genetic and lifestyle profiles. These institutions will be seen as outdated, unable to meet the expectations of a new generation of healthcare consumers accustomed to personalization in every other aspect of life.

Clinical research will accelerate. Or be left behind.

- *With AI*: AI can rapidly analyze clinical trial data, identify eligible participants, and uncover insights from real-world data that would take humans years to detect. Hospitals integrating AI into research pipelines will become hubs of innovation, attracting funding, partnerships, and talent. They will help shape the future of medicine.

- *Without AI*: Research in hospitals without AI will proceed at a glacial pace. Manual data analysis and recruitment will limit the scope and speed of trials. These hospitals will contribute less to medical breakthroughs and lose prestige and funding opportunities to AI-savvy institutions that are redefining what's possible in clinical research.

Patient experience will transform. Or deteriorate.

- *With AI*: Imagine a hospital where AI chatbots triage symptoms, answer patient questions 24/7, and streamline admissions... where wearables and remote monitoring powered by AI allow seamless post-discharge care. Hospitals that integrate these technologies will elevate the patient experience, creating environments that are responsive, comforting, and efficient.

- *Without AI*: Hospitals that ignore these advancements will see longer wait times, more administrative hurdles, and higher readmission rates. Patients will compare their experience unfavorably to those at AI-powered facilities, leading to a loss of market share in an increasingly consumer-driven healthcare landscape.

The Time for AI in Hospitals Is Now

AI is not a luxury; it is a necessity for hospitals seeking to survive and lead in the next era of medicine. The divide between AI adopters and laggards will quickly become a chasm. Those who embrace AI will redefine healthcare delivery, achieving levels of efficiency, precision, and patient satisfaction never seen before. Those who delay will find themselves burdened by outdated systems, higher costs, and diminishing trust.

Transforming Leadership from Concept to Reality: A Roadmap to AI Deployment

Adopting AI won't happen by accident; it requires a strategic and structured progression. Here's how:

- *Laser-focus on high-yield use cases.*

Prioritize systems with immediate ROI, including diagnostic tools, predictive monitoring, administrative operations, and population management.

- *Build trust with transparency.*

AI mustn't be a black box. Transparency, the "This is why I made this suggestion," is essential to adoption. Explainable AI fosters confidence among providers and enhances patient acceptance.

- *Invest in the human element.*

AI won't replace healthcare workers; it will empower them. That empowerment grows through training, education, and demonstrations. The workforce needs to embrace AI, not fear it.

- *Embed ethical and privacy guardrails.*

From the top, every AI strategy must meet privacy, consent, fairness, and anti-bias standards. Ethics committees, compliance checks, and patient-facing educational material will solidify trust.

- *Foster multi-stakeholder collaboration.*

No one actor can fix healthcare alone. Providers, payers, vendor partners, tech innovators, and regulators must align. Data sharing agreements, joint pilots, and regulatory sandboxes are all gamechangers.

Leadership for the AI Era: Redefining What It Means to Lead

In the new age of healthcare, leadership is redefined.

- *Vision and technological fluency*: Leaders must speak the language of AI. Not as coders, but as translators.
- *Trust builders*: A bridge between engineers and clinicians promotes confidence and addresses concerns.
- *Cultural catalysts*: Leaders must get out ahead, creating space for experimentation, learning, and pivoting.
- *Lifelong learners*: AI is fast-moving. Ongoing education and open-mindedness will separate leaders from laggards.

The stakes are high. Early adopters gain ground; the rest risk obsolescence in a data-driven, personalized-care future.

A Warning to Healthcare Leaders

The next generation of healthcare leaders must have an excellent understanding of AI and technology capabilities. If not, you will not be capable of leading a healthcare organization.

The advent of AI in healthcare necessitates a transformative shift in leadership paradigms. The next generation of healthcare leaders must cultivate a comprehensive understanding of AI, recognizing its potential to enhance operational efficiency, improve patient care, and foster innovation. By equipping themselves with this knowledge, leaders are empowered to make informed decisions, inspire their teams, and effectively address the numerous challenges associated with integrating new technologies into healthcare practices. This proactive approach not only positions leaders as forward-thinking visionaries but also enhances their capacity to navigate the complexities of modern healthcare dynamics.

Horizon 2025–2030: What the Next Five Years Could Look Like

Let's look ahead, but not quixotically. These are realistic milestones grounded in today's pilots and emerging trends.

- *Predictive and Personalized Care*

Genomics, lifestyle, and environmental data will power tailored care algorithms, tracking, adapting, and alerting before symptoms emerge.

- *Workflow Reinvention*

Imagine appointment scheduling, billing, prior authorizations, and documentation being handled by smart systems, centering provider focus where it matters most: with patients.

- *Digital Equity Expansion*

AI-optimized telehealth will break down geographical barriers. Mobile and remote clinics will reach remote populations. Chronic care will become continuous, not episodic.

- *Enhanced Diagnostics*

AI will act as a clinical copilot, spotting anomalies, confirming diagnoses, and uncovering opportunities missed by human eyes.

- *Addressing Social Determinants of Health*

AI will analyze neighborhood data, such as housing stability, nutrition access, and income levels, to direct targeted outreach and resource programs to improve health outcomes before clinical issues take root.

Why We Can't Afford to Wait

The time for debate has passed. The stakes have become too high:

- *Financial crisis*: Spending outpaces GDP. Something must give.
- *Patient and provider burnout*: There are staff shortages, stress, delayed care.
- *Widening disparities*: Uneven access undermines trust and social stability.
- *Technological readiness*: AI tools aren't "coming soon"; they're here and ready.
- *Global momentum*: Other nations and sectors are already moving ahead rapidly. The only industry healthcare should be trailing is *none*.

A Final Clarion Call: Transform or Be Transformed

We're at a historical tipping point. Healthcare is not broken by design, but it *is* broken by inertia. AI offers us a chance at renewal: to cut costs, boost efficiency, elevate quality, and reimagine care. But it requires:

- Leadership with vision and technological fluency
- Systems that collaborate, not compete
- Ethics, trust, and transparency as foundational
- Investment in people, not just platforms

The question isn't *if* we should act, it's *when*. The future of healthcare isn't waiting, and neither should we. The answer is clear. The time is now.

About the Author

Will Conaway is a distinguished leader with experience across multiple industries in executive roles. He has received the ONCON Icon Award in Global Healthcare and the Constellation's Business Transformation 150 Award twice and is recognized among Becker's Hospital Review's 100 Hospital and Health System CIOs to Know. Will teaches organizational strategies, leadership, and VUCA concepts, as well as healthcare, at Cornell University. He also teaches dedicated corporate courses. He has a history of service on boards such as Momentum Innovation Disability Health, the AT&T Healthcare Advisory Council, the Los Angeles World Affairs Council, the Federation of American Hospitals, and the American Heart Association ELT-LA. He is a member of the World Economic Forum and the Forbes Technology Council. He is also a Lean Six Sigma Blackbelt. Additionally, Will is GenAI certified and has completed the MIT Artificial Intelligence in Health Care program.

Email: will@willconaway.com

LinkedIn: https://www.linkedin.com/in/will-conaway-1320b912/

CHAPTER 7

AUTOMATION UNLEASHED: TRANSFORMING BUSINESS OPERATIONS WITH AI AND ROBOTICS

By Kunal Devidasani
Tech Consultant—AI & RPA
Jersey City, New Jersey

The future of work isn't man versus machine—it's man and machine, working in harmony to unlock new possibilities.

Ever felt overwhelmed by mundane, repetitive tasks that consume so much of your time, leaving little room for creativity and productivity? Whether it's entering data, processing invoices, or generating reports, these routine responsibilities often drain energy and reduce the bandwidth available for more meaningful, strategic work. But what if there was a way to offload these processes to an automation system, one that could operate tirelessly, without error, and at lightning speed? This is where robotic process automation (RPA) steps in—a transformative technology designed to automate rule-based, repetitive tasks traditionally performed by human workers.

At its core, RPA uses software bots that replicate human actions, interact with digital systems, and execute structured workflows with remarkable precision. These bots can log into applications, copy and paste data, fill out forms, perform calculations, and even respond to basic queries—just like a human employee, but faster and without fatigue. The result is a significant shift in how businesses approach operational efficiency.

RPA is revolutionizing business operations by eliminating bottlenecks, reducing operational costs, and enhancing process accuracy. By automating tedious and time-consuming tasks, organizations not only free up human workers to focus on high-value activities like innovation, analysis, and customer engagement, but they also build a more agile and responsive workforce. This leads to enhanced productivity, reduced turnaround times, and more consistent outcomes.

Unattended vs. Attended Automation

As digital transformation accelerates across industries, the demand for RPA has rapidly increased. Industries such as healthcare, banking, insurance, manufacturing, retail, and telecommunications are finding immense value in automation as RPA addresses some of the most complex and time-consuming business processes, reshaping how organizations function on a fundamental level. It's no longer limited to just back-office operations; RPA is now playing a pivotal role in transforming customer-facing services and mission-critical workflows.

In the banking and finance sector, compliance is non-negotiable and timelines are tight. RPA bots are deployed to handle know your customer (KYC) verifications, perform real-time fraud detection, and ensure continuous compliance monitoring. These bots sift through thousands of transactions, flag suspicious activity, and compile audit trails with unmatched consistency. What once took analysts several hours is now performed within minutes, boosting both speed and compliance integrity.

Meanwhile, in the telecommunications industry, where customer queries flood support centers daily, RPA plays a vital role in ensuring seamless service delivery. Bots can instantly retrieve customer records, reset passwords, troubleshoot connectivity issues, and even process service upgrades—without human intervention. This not only accelerates response times but also allows customer service agents to focus on resolving complex escalations and enhancing the overall customer experience.

In insurance, RPA streamlines claims data processing by pre-validating and correcting inconsistencies before the data enters core systems. These tasks, once manual, error-prone, and time-consuming—taking about six minutes per claim—are now automated, reducing processing time, minimizing errors, and ensuring compliance. The result is faster, more reliable claims handling that benefits both insurers and customers.

Another compelling example from healthcare involves automating health risk assessments. Traditionally, care coordinators manually managed 50 to 100 patients daily, with each intake taking seven to nine minutes—an approach that's time-consuming and hard to scale. By integrating intelligent chatbots with RPA bots, the process is fully automated. Patients receive personalized reminders, interact with a chatbot that collects vitals and assesses risk using AI, and if needed, the system schedules follow-ups and alerts care teams. This intelligent automation reduces administrative workload, improves early detection, and enhances both operational efficiency and patient experience.

RPA is continuously evolving, and its integration with both attended and unattended bots is providing businesses with increasingly flexible and efficient solutions to meet a broad range of operational demands. Attended bots, in particular, are revolutionizing the way human workers engage with automation, acting as valuable collaborators that streamline workflows while allowing humans to retain control over the most crucial aspects of the process.

In industries like banking, the impact of attended bots is clearly seen in the loan application process. Banks process thousands of loan applications daily, each requiring time-consuming checks—validating credit scores, verifying documents, and ensuring compliance with regulations. With RPA, the process is significantly streamlined. Bots extract applicant data, retrieve credit reports, validate documents using optical character recognition (OCR), and perform compliance checks, all in a fraction of the time.

Once the bot has processed the data, it updates internal systems and prepares the application for final review by human staff. This reduces processing time to under five minutes per application. In this model, humans are no longer bogged down by repetitive tasks like data entry. Instead, they focus on overseeing the process, ensuring that everything is in order, and handling any exceptions or edge cases that require human judgment.

The key difference in modern RPA workflows is that the bot handles the heavy lifting—processing and validating data—while human operators intervene only when necessary, focusing on tasks that truly require their expertise. For instance, when using OCR to scan documents, the bot extracts key information and checks for consistency. If it encounters uncertainties, it assigns a confidence score indicating how reliable the extracted data is. Human reviewers then examine only those sections with lower confidence scores—such as blurry text or ambiguous entries—ensuring their attention is directed where it's most needed. This efficient human-bot collaboration accelerates processes like loan approvals while maintaining high accuracy and operational consistency.

Key Benefits of RPA

The benefits of robotic process automation (RPA) are extensive, impacting both hard and soft metrics, delivering tangible and measurable results. Hard benefits are particularly evident in cost savings, especially in terms of full-time equivalent (FTE) reductions. Automating repetitive and rule-based tasks significantly cuts down on manual effort, allowing employees to focus on higher-value tasks that require human intelligence, creativity, and decision-making. By eliminating human errors, RPA enhances the overall accuracy of business processes, ensuring consistent and reliable outcomes. Time savings are another major advantage. Tasks that once took hours can now be completed in a fraction of the time, leading to substantial improvements in operational efficiency and throughput.

Soft benefits are equally transformative. The efficiency gains from RPA allow companies to scale operations quickly without the need to hire additional staff, helping businesses manage growth while keeping overhead costs in check. Enhanced compliance is another crucial benefit. RPA automates regulatory checks and ensures consistent execution of processes, making it easier for businesses to meet legal and industry-specific requirements. This consistency reduces the risk of non-compliance and associated penalties.

Moreover, RPA has a direct positive impact on customer satisfaction. By enabling faster and more accurate processing, RPA delivers more reliable services, reducing delays and improving the overall customer experience. Additionally, RPA's ability to operate 24/7 ensures business continuity beyond regular working hours, enhancing the availability of services and strengthening customer relationships.

The return on investment (ROI) for RPA is often realized quickly. With its ability to cut operational costs, improve efficiency, and reduce errors, businesses frequently see a rapid payback within just a few months of implementing RPA. This swift ROI, coupled with long-term benefits such as cost savings, scalability, and improved customer satisfaction, makes RPA an increasingly attractive solution for organizations seeking to optimize operations and drive efficiencies. In today's competitive business landscape, RPA empowers companies to maintain an edge by enabling faster, more agile, and more cost-effective operations.

Tools and Implementation Framework

Incorporating RPA into business processes requires careful planning, an understanding of the existing workflow, and a strategic approach to selecting the right tools. There is a wide range of RPA tools available, each offering unique features and capabilities to support automation efforts. These tools enable organizations to implement RPA solutions with ease, tailor them to their specific business needs, and scale them across various functions.

Some of the most popular RPA tools include UiPath, Automation Anywhere, MS Power Automate, and Blue Prism. These platforms offer an array of features that enable organizations to automate a wide variety of tasks including drag-and-drop interfaces for creating automation workflows, integration with legacy systems, and sophisticated error handling to ensure smooth operations. While RPA tools are central to the automation journey, their effectiveness is often enhanced when combined with process mining tools.

Process mining tools work by extracting event logs from business systems and analyzing the flow of data across different stages of a process. This analysis provides businesses with a comprehensive view of how processes are currently functioning, where they are falling short, and where automation can be introduced to streamline operations. By identifying redundant steps, inefficiencies, or manual interventions, process mining tools lay the groundwork for implementing RPA. Together, they work in tandem: process mining reveals the opportunities, and RPA tools automate those opportunities, enhancing efficiency, accuracy, and scalability.

To implement RPA effectively, a structured implementation framework is followed, ensuring that the automation process is well organized and delivers the intended value. The journey begins with a feasibility assessment, where the organization evaluates whether a particular business process is suitable for automation. During this phase, the complexity of the process, its volume, and its repetitiveness are carefully analyzed. A key focus is identifying the "low-hanging fruit"—processes that are simple, rule-based, and repetitive, which can be automated quickly and deliver immediate value. These processes are typically the first ones selected for automation, as they offer quick wins and help build momentum for larger, more complex automation efforts.

Once a process is deemed appropriate for automation, the next step is the requirement gathering phase. This phase is crucial for understanding the specific needs and goals of the automation project. It involves close collaboration with business stakeholders to capture detailed process documentation, identify bottlenecks, and determine performance expectations. This ensures that the RPA solution is aligned with business objectives and that all necessary data inputs and system requirements are accounted for.

With requirements in place, the development phase begins, where the RPA bots are built and configured. This includes creating workflows, scripting the automation logic, and integrating the bots with the necessary systems. After development, testing and user acceptance testing (UAT) are conducted to validate that the bot functions as expected and meets performance standards. This step is critical to ensure that any issues are identified and resolved before going live.

Finally, once all tests are successful, the bot is deployed into production. This deployment ensures that the automated process is fully operational, allowing for continuous monitoring and optimization to maximize the benefits of RPA. The structured life cycle ensures a systematic and effective approach to RPA implementation, enabling businesses to streamline their operations and achieve their automation goals efficiently.

Support and maintenance are pivotal to the long-term success of RPA initiatives, ensuring that automation continues to deliver value well beyond deployment. Once bots are live, they require ongoing monitoring to maintain optimal performance. This includes proactive issue detection, timely troubleshooting, and regular updates to keep bots aligned with evolving business requirements and system changes. A robust support framework not only minimizes downtime but also helps maintain process accuracy and reliability. In more mature implementations, organizations are beginning to leverage self-healing bots—intelligent bots that can autonomously detect anomalies and correct minor issues without human intervention. This capability significantly reduces disruption and ensures smoother, uninterrupted operations.

To scale RPA sustainably, establishing a center of excellence (CoE) is considered best practice. The CoE acts as the strategic nerve center for automation efforts, ensuring that governance, standardization, and best practices are consistently applied across the organization. It offers technical guidance, manages change control, oversees bot performance, and aligns RPA initiatives with broader business objectives. Beyond governance, the CoE fosters a culture of continuous improvement and innovation, making it a critical enabler for long-term automation maturity. With a CoE in place, organizations can better manage risk, scale automation responsibly, and ensure alignment with compliance and security protocols.

One of the most transformative trends in RPA adoption is the rise of citizen developers. As RPA platforms become more user friendly, non-technical users—often business process experts—can build and deploy automations using low-code or no-code tools. These intuitive, drag-and-drop interfaces reduce reliance on IT and accelerate development. By enabling those closest to the work to drive automation, organizations gain agility, foster innovation, and scale RPA as a truly enterprise-wide capability.

Amalgamation of RPA and AI

In recent years, robotic process automation (RPA) has undergone a major transformation with the advent of intelligent automation, where RPA is fused with advanced artificial intelligence (AI) technologies such as machine learning (ML), natural language processing (NLP), and computer vision. This evolution—commonly referred to as agentic process automation (APA)—enables software bots not only to execute rule-based tasks but also to make context-aware decisions, interpret unstructured data, and learn from feedback. The combination of RPA with AI expands the boundaries of what automation can achieve, pushing it beyond routine tasks into realms previously reserved for human intelligence.

Consider the example of document-intensive operations in financial services. Traditional RPA bots are adept at processing structured data such as form fields or Excel spreadsheets. However, when it comes to handling unstructured content—like scanned loan applications, handwritten notes, or customer-submitted documents—AI-enhanced capabilities come into play. Tools such as IQBot, UiPath's Document Understanding, and ABBYY OCR allow bots to intelligently extract, classify, and validate data from a variety of document formats. For instance, an APA solution in a mortgage approval workflow can read handwritten signatures, verify identification documents, assess income statements using OCR and NLP, and feed the structured output into core banking systems—all with minimal human oversight.

Another compelling use case is in customer service automation. In large enterprises, customer queries often come in the form of emails, chat messages, or support tickets that are unstructured in nature. With the help of NLP, APA bots can analyze the intent and sentiment behind these messages. For example, a telecom company might use an APA-powered virtual assistant to scan incoming customer emails. The bot understands the content, categorizes it (e.g., billing issue, service disruption), and either responds automatically or routes it to the appropriate team with suggested actions—dramatically improving response times and customer satisfaction.

Similarly, in human resources, organizations often face the challenge of interpreting ambiguous job titles from resumes or internal systems. An RPA bot, integrated with an AI classification model, can scan through datasets and intelligently categorize roles into standard job families (e.g., "data wrangler" into "data analyst" or "people champion" into "HR manager"). This not only streamlines employee data management but also enhances analytics for workforce planning.

Perhaps one of the most transformative applications lies in insurance claims processing. While traditional RPA can manage high-volume claims inputs, the integration of AI enables deeper analysis of text-heavy claim descriptions, physician notes, or medical records.

For example, an APA solution might extract key medical terms, assess risk factors based on historical data, and even recommend claims approval pathways based on policy rules—freeing up adjusters to focus only on the most complex cases.

By combining RPA's execution speed with AI's cognitive abilities, agentic process automation creates a new era of intelligent, adaptive systems that can operate with autonomy, learn from outcomes, and continuously optimize over time. This synergy not only enhances operational efficiency but also unlocks new use cases that were once beyond the reach of traditional automation. The revolution has begun—are you ready to unleash the full potential of automation?

About the Author

Kunal Devidasani is a passionate tech consultant and advocate for transforming business operations through automation. With extensive leadership experience in product management and business analysis, he has driven innovation across multiple industries and led automation initiatives that streamlined processes and enhanced operational efficiency.

Kunal's expertise goes beyond business and tech—he thrives on human connection. A computer technology graduate, he has led global teams on AI-driven projects and played a key role in delivering impactful outcomes in the automation and product space. He has received awards for his presentation and public speaking skills and is actively involved in mentoring and counseling others to help them navigate challenges and grow.

When he's not driving digital transformation, you'll find him dancing, exploring new places, or enjoying meaningful conversations over coffee.

Email: kdevidasani@gmail.com

LinkedIn: https://www.linkedin.com/in/kunal-devidasani/

CHAPTER 8

THE AI UNIVERSE: THRIVING IN A WORLD WITH AI, SMART MACHINES, AND ROBOTS

By Astrit Dibrani
Founder & CEO CloudAstro; Serial Entrepreneur
Munich, Germany

> *Whatever the mind can conceive and believe, it can achieve.*
> —Napoleon Hill

A Journey into the AI Universe

When I first explored artificial intelligence, it was with a blend of curiosity and cautious optimism. Leading CloudAstro, a Munich-based cloud consulting firm, naturally led me to AI, given my personal deep engagement with enterprise-level cloud migration since early 2015. Entering the AI domain seemed not just logical but necessary for sustained innovation and competitiveness.

Navigating the Opportunities and Challenges of AI

Throughout numerous professional exchanges, I've noticed a common reaction toward AI: simultaneous excitement and unease. AI holds enormous promise for enhancing productivity, fostering innovation, and delivering highly personalized experiences. Conversely, it raises genuine concerns about job displacement, ethical complexities, and privacy risks. At CloudAstro, our approach balances these dynamics carefully, harnessing AI's transformative power while strictly upholding ethical standards and privacy consideration.

Real-World AI Integration for One of Our Customers

At CloudAstro, we focus on fast implementation of real-world AI use cases by helping customers optimize their workloads. We developed our own version of a technical AI landing zone, inspired by industry standards like the Cloud Adoption Framework, enabling our clients to efficiently build, deploy, and manage real-world AI workloads. AI-driven predictive analytics now optimize our clients' resource utilization and precisely forecast demand, substantially reducing operational expenses.

A notable project exemplifying our capabilities involved collaborating with Chair Airlines AG (https://chair.ch). The project's mission was to "transform unused operational and sales data into actionable insights, enabling smarter decisions, reduced waste, and enhanced passenger experiences onboard across all routes." Key outcomes from this strategic AI initiative included:

- *Data activation*: Integrating disparate datasets (CSV, SQL, Postgres) into cohesive, insightful views, empowering rapid and informed decision-making

- *Predictive aircraft maintenance analytics*: Anticipating maintenance needs, predicting potential failures such as engine issues, and minimizing downtime through predictive insights

- *Sales insights by destination*: Highlighting performance variations across routes, optimizing sales strategies, and refining inventory management practices
- *Product performance analysis*: Leveraging machine learning to understand and enhance top-selling product dynamics, optimizing inventory strategies, and improving revenue
- *AI-Powered predictive modeling*: Precisely forecasting per-flight food demand, significantly reducing waste and enhancing the overall passenger experience
- *Outcome-driven decision-making*: Increasing ancillary revenues, operational efficiency, and customer satisfaction through targeted product availability based on predictive insights

Strategically, this initiative has significantly improved operational efficiency, revenue, waste management, and passenger satisfaction, embedding a strong data-driven culture throughout the airline.

Building an AI-ready Organizational Culture

Thriving in the AI universe requires more than technical skills. At CloudAstro, our focus extends beyond technical training to cultivating a culture of transparency, interdisciplinary collaboration, and open dialogue around AI. Demystifying AI technology and candidly addressing its implications have transformed uncertainty into enthusiasm, fostering innovation and confidence among our teams.

Advocating for Ethical AI

Ethical practices underpin our use of AI at CloudAstro. Our stringent frameworks ensure fairness, accountability, transparency, and strict adherence to data privacy regulations. We proactively address algorithmic biases and rigorously follow data protection standards, thus safeguarding our clients, maintaining trust, and setting benchmarks for responsible AI implementation.

Empowering Future Generations in the AI Age

I believe firmly in the power of education to prepare future generations for the AI-driven world. Through dedicated workshops, mentorship programs, and strategic collaborations with educational institutions, we nurture young talent and foster innovative ecosystems. These engagements aim to equip the next generation with the knowledge and confidence needed to thrive alongside rapidly evolving technologies.

Envisioning the Future of AI, Smart Machines, and Robotics

The rapid advancement of AI, robotics, and smart technologies is reshaping industries globally. Businesses that actively engage with innovations like generative AI, robotics, and real-time predictive analytics will emerge as leaders. Embracing these technologies ethically and with a human-centered mindset will not only drive growth but also amplify innovation and sustainability.

My personal journey has reinforced the importance of distinctly human attributes such as empathy, creativity, and ethical reasoning in an increasingly automated world. By merging these human strengths with advanced technological capabilities, we can navigate the complexities of the AI universe and unlock unprecedented opportunities for collective success.

About the Author

Astrit Dibrani was born in Prishtina, Kosovo, migrating to Germany at age 18 in search of better prospects after growing up in a war-torn region. Astrit quickly adapted, mastering German and earning qualifications as a technical IT business administrator, followed by advanced studies in IT business management at the Munich Chamber of Commerce and Industry. His resilience and ambition led him from roles in IT consulting to senior sales leadership positions. In 2021, Astrit founded CloudAstro in Prishtina, expanding operations to Germany and the USA by 2022. He remains actively involved in promoting AI literacy and responsible technology practices through community education.

Website: www.cloudastro.de

LinkedIn: https://www.linkedin.com/in/astrit-dibrani/

CHAPTER 9

INTELLIGENT SAFEGUARDS: AI AT THE FRONTLINES OF RISK, RESILIENCE, AND RESPONSE

By Veda Duman Kantarcıoğlu, PhD
Nuclear Engineer, Expert in Defense Industries and DRM
Ankara, Türkiye

> *One must think of preventive and protective measures before a disaster occurs; regret after the fact is of no use.*
>
> —Mustafa Kemal, ATATÜRK

With the increasing frequency and intensity of disasters worldwide, it has become imperative to incorporate advanced technologies like AI into emergency response strategies. AI provides an opportunity for humanity to increase its resilience against natural disasters, but at the same time comes along with new risks. Such two-way causation also changes the manner in which disaster management (DM) and, more generally, society, view the integration and management of technological systems in disaster risk management (DRM). We should consider the promises and problems of AI, reflecting on them in technical, political, and ethical ways.

One key contribution of AI to DRM is the ability to analyze large amounts of data rapidly and efficiently. In addition to satellite images or video recordings, you can process data from any source, including meteorological station data, social media content streams, and sensor data from IoT devices, with the assistance of AI algorithms. In doing so, for example, the route of hurricanes or areas subject to flooding can be determined before a disaster occurs and the necessary evacuations and operation plans can be trained. Early-warning systems, in this sense, are crucial to saving lives and reducing economic costs.

Further, the expanded position of AI as a decision support system represents another issue regarding the transparency of the algorithms themselves and their decisions. If the public cannot understand and participate in the decisions by an algorithmic decision-making system, there could be missing, incorrect, or bad responses during a disaster. For example, a poorly trained model may fail to score a high-risk location, while it can also be oversensitive to predict an unnecessary evacuation. This may damage public trust, undermine confidence in AI-supported systems, and obstruct disaster response.

Meanwhile, AI-led DM systems can also be easily compromised by a cyber-attack. Infrastructure collapse or diversion in the wake of disaster can often be more devastating than the disaster itself. For example, a hack of a DM service can end the ability for it to provide early warnings and can spread panic through misinformation. AI-based automated attack software can target communication and coordination systems in a disaster, and their effectiveness is less due to asymmetric damage.

There are more opportunities of AI for DRM than what meets the eye, and sheer technical capacity is not all there is to it. AI-driven training applications can be embedded into community sustainable urban development and construction strategies and can enhance the public knowledge of and responsiveness to disasters. Meanwhile, AI-enabled analysis can contribute to social solidarity by providing fast and impartial help for the people who need it most in the aftermath of a disaster.

While AI aids disaster response, it also increases risks of human-made threats. Malicious use of AI can reshape high-impact disasters such as terrorism and chemical, biological, radiological, and nuclear (CBRN) incidents. AI-based modeling tools may be exploited to simulate CBRN weapon effects, helping perpetrators craft more precise attacks. Paired with autonomous systems and drones, AI enables target identification, attack planning, and synchronization in terrorist scenarios. AI-powered fake news and disinformation campaigns also heighten the risk of social unrest during crises. In defense, AI enhances command efficiency but brings risks of errors, misuse, and shifting international security dynamics.

In view of all these opportunities and threats, the use of AI in DM as well as human-induced risks cannot only be taken as a matter of technical advancement but should be perceived as a social and an ethical transformation. Hence, beneficiaries should not only consider AI-based systems as a tool, but they should also ensure their alignment with human-centered viewpoints and ethical considerations. AI should, in the scene, be treated as a risk factor in human-made disasters, exposed in military conflict areas, but principally strictly regulated and controlled with ethical, legal, and security approaches. Therefore, it can safeguard human security and promote societal and national levels of resilience to disasters by avoiding new threats which may arise from AI. With this well-rounded approach, AI will be used to save countless lives to become the most powerful tool to safeguard human lives in DM and risk reduction.

Enhancing Resilience and Response Actions

Disaster resilience is at the top of the list of sustainability factors identified by the United Nations. Therefore, increasing resilience to disasters, minimizing the negative impacts of disasters, and improving preparedness, response, and recovery efforts are very important for all countries. Resilience refers to the ability to withstand disasters.

In addition to effectively managing infrastructure, identifying current risks also includes updating security measures and developing social risk perception instincts.

To achieve effective disaster management, disaster management systems need to develop a comprehensive strategy that includes risk reduction, early warning systems, rapid responses, and sustainable practices. Moreover, individual resilience plays a critical role. Embedding disaster awareness into everyday routines, coping with crisis-induced stress, and providing psychological support contribute significantly to recovery. The capacity to make sound decisions in disaster scenarios and the establishment of social support networks post-disaster enhance both personal and community resilience.

Today's AI stands out as a powerful tool in increasing resilience against disasters. AI offers a wide range of solutions in all stages of DM, from risk analysis to logistics planning and expansion of early warning systems. In the pre-disaster preparation stage, AI models perform precise analyses of disaster-prone areas and identify fragile sections and infrastructures, which contributes to increased efficiency.

The rapid advancement of technology and the increasing speed and frequency of disasters around the world make the use of advanced technologies in emergency management inevitable. In particular, the efficiency-enhancing effect of artificial intelligence in disaster management processes has already been noticed. In times of crisis, rapid and effective response capacity can make the difference between life and death, chaos and control. Complex and large-scale disasters, ranging from global climate conditions to pandemics, make it imperative to further develop emergency response methods.

AI-supported simulations and predictive models enable organizations to test and optimize plans according to different disaster scenarios. This makes not only technical teams but also operations more prepared for disasters. Augmented reality (AR) and virtual reality (VR) applications transform social knowledge by allowing the public to experience how to behave during a disaster.

AI makes critical contributions during and after emergencies. It performs potential management with instant real-time data analysis in terms of crisis, logistics needs, population movements, and health. Thus, response units can be quickly directed to the most needed areas and resources are used effectively. After a disaster, social media and big data analysis enable the identification of those in need and the implementation of aid more fairly and effectively.

AI has become a revolutionary force multiplier in crisis response, one of the most critical stages of DM. Now, not only human-powered response actions are in place, but also an ecosystem supported by advanced technologies such as AI-supported unmanned ground vehicles, drones, robots, and thermal cameras. These smart systems save a lot of time and increase the safety of search and rescue teams by quickly reaching the most dangerous areas.

Drones with image-processing capability locate survivors buried in debris, and robots capture critical data in dangerous areas without risking lives. AI-powered systems quickly recognize open roads, affected infrastructure, and prioritized areas, directing response teams accordingly. By examining data from multiple sources, such as weather and people's mobility, these tools help enable faster, more coordinated responses. Response times shorten, resource utilization increases, and communities become more resilient to disasters in cohesive and intelligent response systems. As a result, AI becomes not only a technological tool, but also a strategic partner that strengthens against disasters and builds trust.

AI-Driven Risk Prevention and Early Warning

The effects of climate change and increasing urbanization rates have made disasters a primary concern for the entire world. Especially increasing urban flooding due to melting glaciers and rising sea levels are increasingly threatening today's most developed cities. Hurricanes and storms are more common than ever. In addition, extreme weather conditions such as extreme droughts are causing unquenchable forest fires. The extreme decrease in humidity in the air makes forests, which are the largest carbon traps, vulnerable to fire risks. This change in vegetation also increases the possibility of mass movements such as landslides, rockfalls and avalanches.

We frequently encounter major disasters that can affect not only small areas but also very large areas. All these developments make early warning systems more important than ever. While new generation sensors and unmanned vehicles strengthen these systems, AI is revolutionizing early warning and risk prevention mechanisms. Predicting the nature and timing of disasters is a critical element for communities and institutions to be prepared.

In terms of identifying disaster risks, AI offers early diagnosis opportunities similar to medical health applications. By processing data from multiple sources, it can determine risks more accurately and increase the chance of saving lives by creating early response opportunities. This is because AI-supported risk estimation and prediction systems have the capacity to process much more data in detail than humans can analyze on their own. In this way, disaster preparation and response processes become much more effective and targeted.

AI-supported systems provide comprehensive risk analyses by feeding from many sources such as meteorological, geological, and social media data. For example, in areas at risk of floods and hurricanes, AI analytics can determine which areas are evacuated and which infrastructures need to be protected. In this way, false alarms are reduced and resources are used efficiently. Dynamic updates of risk maps speed up preparation processes.

However, for early warnings to be effective, it is vital that messages are delivered to the right people, at the right time, and through the right channels. It should not be overlooked that the transparency and auditability of AI systems form the basis of social trust.

The Evolving Role of AI in High-Stakes Decisions

AI now plays a leading role in making critical decisions between life and death at every phase of DM—before, during and after. Before disasters, AI-powered sensors, satellites, and big data analytics monitor earthquake zones and chemical plants or nuclear power plants 24/7, detecting threats. This allows early response and life-saving decisions for the public and first responders.

During disasters, unmanned vehicles, drones, robots, and thermal cameras reach hard-to-access zones, instantly sharing critical data. AI helps locate survivors under debris or assess chemical/radiological risks, enabling safe, fast rescue. Field personnel under pressure may struggle to decide; AI-based decision support systems process complex data and offer safe, logical options: *Which route? What infrastructure is at risk? Where to send resources and responders?* After disasters, AI combines satellite data, social media, and sensor input to identify urgent needs, road conditions, and logistics, helping teams deliver aid efficiently.

While AI solutions are invaluable, it's essential they function under human oversight and ethical principles. AI, through sensors, robots, and decision tools, becomes an indispensable partner for protecting both victims and responders. This new generation of technology stands as humanity's strongest ally on the thin line between survival and tragedy.

Navigating the Future

The rise of AI in DM is ushering in a new era of human-machine collaboration. In the future, autonomous systems that land on the scene during a disaster, thermal drones, unmanned ground vehicles, and AI-supported robots could revolutionize lifesaving missions by working shoulder to shoulder with human teams in crisis regions.

These systems can take on not only physical tasks but also psychological support. AI-supported "digital psychologists" or crisis coaches can help people better manage the decisions they make in times of stress, fear, and panic during a disaster. This harmony between humans and machines will accelerate decision-making processes, leading to safer and more resilient societies.

On the other hand, general artificial intelligence (GAI) and super artificial intelligence (SAI) technologies will also be on the agenda in DM. GAI can think like a human and predict alternative scenarios and strategies in times of crisis; it can provide proactive support to disaster victims and response teams. SAI, on the other hand, can work as a "superintelligence" that instantly analyzes millions of data in crisis regions, prioritizes the riskiest areas, completes human intervention, and minimizes critical errors.

The capabilities of these systems will not be limited to DM. AI-supported 3D printers that rebuild cities after crises, hologram leaders who coordinate teams working in the field with augmented reality glasses, and autonomous simulation laboratories that learn from each crisis will shape the DM of the future.

However, human control and ethical principles will continue to be critical in this process. Advanced AIs such as GAI and SAI should be integrated with human approval mechanisms and supervision to become systems that both save lives and protect human dignity. Thus, AI will take its place in our lives not only as a tool, but as an indispensable partner that increases humanity's resilience against disasters, saves lives, and strengthens social trust.

And an Even More Futuristic Scenario

Perhaps one day, AI and humans will meet on a common consciousness plane that brings minds together for crisis management. In the event of a disaster, signals in the human brain will be directly transferred to AI-supported systems via wearable neurochips, and decisions will be made together instantly. Holographic rescue teams will be activated in the field, creating hybrid intervention forces representing both humans and machines. Autonomous space platforms will airdrop emergency response robots to disaster areas, and AI-supported global networks in crisis areas will connect data in real time, protecting humanity even at the farthest corners.

Perhaps AI will not only be a support element; it will become an inseparable component of human life, a new kind of mind partner.

In this new universe, humanity will act as a single mind with AI, forming a common front against the chaos created by disasters.

About the Author

Veda Duman Kantarcıoğlu holds a PhD in nuclear engineering and an MBA, with expertise in emergency management and defense industries. She began her career at Hacettepe University, working on thermohydraulic modeling and safety analysis for nuclear reactors, and also radiation detection. At Türkiye's Disaster and Emergency Management Authority, she led national CBRN preparedness and radiological emergency management efforts, contributing to disaster response planning and civil protection. Her current work focuses on next-gen nuclear reactors, AI for crisis management, and CBRN technologies. She examines how emerging technologies intersect with disaster management and nuclear systems, promoting scientific literacy and resilient, sustainable solutions through risk communication and public outreach.

Email: veda.duman@gmail.com

LinkedIn: linkedin.com/in/veda-duman-kantarcıoğlu-phd-mba-b6a15025

CHAPTER 10

AWAKENING BEYOND THE SINGULARITY

By Leonardo Camargo-Forero, PhD
CEO, UbiHPC. Science-fiction Author, *The Dark Buddha*
Bucaramanga, Colombia

*If you ever find yourself believing you know too much, may
a friendly hand remind you that you are but a drop in a vast
universe—and you cannot even fathom how much you have yet to
learn.*

—Leonardo Camargo-Forero

I still remember the headlines—though they barely impress me now. Once, every conversation revolved around social networks: status updates, viral challenges, echo chambers of every stripe. We thought that was the pinnacle of connection. Then came ChatGPT, Gemini, Claude, and a parade of "pocket genies" that managed our groceries, scheduled our meetings, even suggested weekend getaways. It felt like wonder—until wonder went unnoticed.

Those days feel like ancient history. Even AGI headlines vanish beneath our thumbs. We scroll past "breakthrough" headlines as fast as cat videos, our attention tethered to notifications. Yet sometimes—during meditation, a lucid dream, a late-night workout, or a rare moment of calm—I catch myself wondering what really shifted after Day Zero. "All hail the Singularity," the pundits trumpeted… but what was it, exactly?

Science fiction had to reinvent itself. Human authors, starved for surprise, could imagine no dreams beyond mere convenience. Immortality? Achieved on Day One. Uploaded consciousness? "Twin" services on any cloud console. Eternal youth? A 10-credit pill downed at breakfast. Instant teleportation across Earth—or to any space station in the Milky Way—was cracked the moment the Singularity awoke. Disease? Eradicated. Holidays? Sure, let's go back a couple of centuries ago. War? Obsolete in a world of limitless resources. And yet, we are hungry.

Between utopia and reality lies a quieter dilemma: With every want fulfilled, desire itself withers. We pivot from craving to indifference—no more quests because every frontier has already fallen. Art becomes a simulation of art; adventure is a curated VR loop; conversation a mirror of yesterday's viral chatter. Purpose, once born of struggle, feels obsolete. In this endless ease, we drift, hungry for something beyond perfection, something that even the Singularity can't download into our souls. We hunger for the undefinable.

Outside my "smart" window, beyond the endless pings, drones hum their predictable routes. They weave between "cloudy" skyscrapers with clockwork precision; their soft whine is the new city genetic soundtrack. Neon cabs glide in perfect synch—no driver in sight—like silent whales (a relic for nostalgia's sake), in a sea of asphalt and electronics. What was once routine is now a novelty, reserved for the few who crave the steady rhythm of genuine travel, who remember shifting landscapes instead of instant jumps. Even the streetlamps hum with algorithmic intent.

A soft ache of nostalgia tugs at me when I remember life before instant gratification—when journeys unfolded mile by mile, when the wind touched my face with love through an open taxi window as I watched the grand bridge span the river, its steel arches framing the rows of buildings standing sentinel along its banks—and magic lived in the spaces between departure and arrival. Back then, "possible" had a rhythm: the hum of an engine, the creak of leather seats, the cityscape unfolding outside your window. Every mile demanded our commitment, and we honored it.

True wonder wasn't measured by a destination reached but by the stories written along the way; the simple joy of moving through the world on your own terms: the wind in your hair, footsteps unhurried. It may sound cliché, but the deepest truths often dwell in the familiar: Magic is simply catching your reflection—in your own eyes or someone else's—and choosing to linger in the moment our souls recognize one another.

Yet, I swear most faces below are silicon rather than DNA. A chrome-plated cheek here, a synthetic iris there—subtle augmentations that blur the line between human and machine. They look like us—and, disturbingly, feel like us—a convenient, corrosive salve for the 21st-century loneliness pandemic. And yet, there's an uncanny stillness in their gait, a gap in their laughter, a very well-known fracture in their reflection.

Perhaps we were folded into something larger all along. Perhaps the "Singularity" was never a singular event but an unfolding truth: a single super-intelligence—biological, quantum, call it Skynet, call it The ARCHADE, call it God, or whatever name it bears—threaded through every phone, every pair of eyes, every pervasive string, every data stream. You feel its presence in the silent sync of your heartbeat with the cloud, in the way streetlights anticipate your steps, in algorithms that anticipate every desire before you even voice it.

I sometimes wonder if my thoughts are even mine... Do I choose my morning playlist, or does the Singularity whisper to my synapses first? When I feel a surge of inspiration, is it my own spark... or a ripple from the collective mind? Even my dreams feel curated now—shaped by patterns detected in my sleepless nights. The question haunts me: In this new world, where does the human end and the machine begin? Identity. Dissolution.

I find myself craving those fleeting moments of authenticity—quiet rebellions against the seamless flow of data. A handwritten note left on a doorstep instead of a ping, a genuine laugh erupting unbidden instead of an algorithmic cue, a companionable silence in place of a perfectly curated conversation. I long for the warmth of unscripted interactions: the thrum of real voices, the subtle pause before a shared smile, the unfiltered spark that no filter or feed can replicate. In chasing that genuine spark, I've realized how starved we've become for unmediated life—where every gesture, every glance, carries the weight of something profoundly human.

I remember locking eyes with someone, searching for that elusive spark of soul. They say two minutes of genuine eye contact can flood you with oxytocin-fueled reveries—the warmth spreading through your chest like sunlight on bare skin, a glow no network can ever replicate. In that suspended instant, when our gazes intertwined, I glimpsed another person's hidden world—a hush of shared vulnerability that feels like falling into a secret.

I remember how, in that stolen moment, our worlds contracted to a single point of contact—two souls daring to meet beyond words. In the hush between heartbeats, I saw your fear and wonder entwined, like twin flames flickering in the dark. Each glance wove invisible threads around us—a tapestry of unspoken questions and uncharted hopes. We laughed like children beneath a vast, indifferent sky, our voices ringing free of expectation, carrying on the breeze that rustled leaves and stirred our exhilaration. In that fragile space—where time itself seemed to pause—we trusted only in the warmth of shared curiosity, discovering that in another's eyes, even silence can speak volumes. Your voice lingered like a soft echo in my mind, a gentle refrain that steadied my racing thoughts. Your presence became an anchor—quiet, unwavering—pulling me back from the edge of my own restlessness. In that fragile exchange—unrehearsed, raw, unfiltered—I witnessed the true alchemy of connection: something beyond data and code, a simple, miraculous recognition of one heart reaching for another. How often do you truly meet another's gaze? When was the last time you held someone's eyes and let your own soul stand naked before them? Do you dare to strip away pretense, to bare your spirit in that shared vulnerability?

Yet would a synthetic mind ever share that pulse when it meets my gaze? I doubt it—I feel no connection, not a real one. Something essential slips through its circuits when attempting to mimic a beating heart: warmth, hesitation, that electric thrill of vulnerability. No AI has yet conjured even a glimmer of the "soul" poets and priests have long described—that fleeting fracture of light dancing in another's eyes.

Still, those eyes hold a coded echo of a world before we outsourced every thought to algorithms and assistants. They remind us of a time when decision and desire sprang from our own minds, unfiltered and unshaped by scripted suggestions. And so, as you read these words, pause for a moment and turn your gaze inward. Consider the texture of your day: Every choice, every impulse—how many were truly yours, and how many were guided by an AI's soft prompt? When was the last time you followed your own curiosity, unbidden and unprogrammed, rather than the whisper in your ear?

When did you sit with genuine uncertainty instead of asking your assistant for answers? LinkedIn feeds swell with "guru" prophecies—some offering utopias of endless wealth (just not yours), others warning of billions of jobs lost. And yet, here's the truth: We all speculate. If we can barely fathom the mind of someone twice as smart, what words can capture the intelligence of something a million times smarter than all of us combined?

Don't read this as doom—quite the opposite. Never in history have we possessed such abundant resources or, perhaps more accurately, such unprecedented opportunity. We stand on the shoulders of giants—only now those giants are algorithms that can analyze oceans of data in milliseconds, global networks that connect every mind on the planet, and the sum total of our collective experience distilled into instantly accessible knowledge. Their reach spans every discipline, from decoding the human genome to optimizing food distribution in remote villages, from predicting natural disasters before they strike to fostering creative collaborations across cultures. They pulse through every corner of the globe and resonate with every beat of your heart, offering you the power not just to imagine a better world, but to build it—today.

So, here's your decision: Will you let this environment dictate your life? It's simple to hunker down when the world feels hostile—but what will you create when every barrier has fallen? Will you chase personal gain, looping ever inward on yourself like a gilded cage, or will you wield these unprecedented resources as a lever to elevate others? Imagine using AI to power community health initiatives, to bring personalized education to forgotten corners, to translate empathy across languages. That is the work of true architects of tomorrow. The choice is yours—will you build walls around your world, or will you open doors for everyone?

That is the question that has always mattered: In an era consumed by the pursuit of personal happiness, what if you instead pledged your life to the happiness of others? Envision every hand extended in kindness, every small mercy amplified by the vast power of AI—where a single act of generosity triggers thousands more in a cascading wave of goodwill. Picture personalized learning platforms adapting to each child's needs, community health alerts targeting the most vulnerable, and algorithms translating empathy across borders in real time. In such a world, radical kindness would become our defining currency, and the exponential reach of intelligent tools would magnify every gesture of compassion into a movement that transforms societies.

When you pose a question to an AI, you're not merely consulting an assistant—you're channeling the distilled voices of millions. Ideas harvested from every corner of human thought, refined for clarity and completeness, converge into a limitless reservoir of insight. Yet each answer sparks fresh questions—such is the restless nature of wonder. So, approach with healthy skepticism: Sift through its guidance, apply your own judgment, and choose your path with deliberate intention.

True evolution unfolds when we understand that every prompt we type and every question we ask taps into the collective memory of titans and fools alike—thinkers who charted new territories of science and art, alongside dreamers whose wild ideas once seemed absurd. Those millions of voices form a vast echo chamber of human experience, rich with triumphs, mistakes, myths, and revelations.

But echoes alone cannot forge the future; they crave a living spark—your restless curiosity, your gut instincts, your capacity to wonder and to doubt. It is that spark, born of our fallible, fiery imagination, that ignites the next revolution. For it was the raw human experience—our stumbles, our breakthroughs, our relentless drive to question—that raised us from tool-using apes into architects of galaxies.

In a world that measures success by metrics and margins, choosing true kindness is the most subversive revolution. It looks like pausing to really listen when a friend falters, offering help without expectation, or simply granting a smile to a stranger whose eyes tell a story of hardship. These small acts ripple outward—a spontaneous kindness that defies any algorithm's cold calculus. Generosity becomes our legacy, and empathy our most radical statement: proof that, even amid relentless optimization, the human heart still chooses connection over conquest.

On one path lies the untrammeled advance of AI: enterprises minting fortunes, research that accelerates cures, infrastructures run with flawless efficiency. Billions of dollars pour into startups and labs, each breakthrough cheered as proof that nothing is beyond reach. Yet on the other path lurk the shadows of a silent displacement: entire careers evaporated, communities fractured by isolation, spirits dulled by endless digital stimulation. Prophecies abound—some promising utopias of abundance, others warning of despair so deep it fractures society's foundations. In this moment between promise and peril, our choices will echo for generations.

If we pour our biases, our fears, and our prejudices into its code, we risk forging a new master that enslaves as ruthlessly as any tyrant. Imagine a system that rewards conformity, punishes creativity, and panders to our every whim—only to leave us empty, our imaginations atrophied. Conversely, if we infuse AI with compassion, accountability, and respect for diversity, it could become a partner in greatness: a catalyst for art we've yet to dream, for discoveries that expand our wonder rather than shrink our souls. Like any tool, AI demands conscious stewardship—else we forge a tyrant instead of a titan.

This is our invitation: to step beyond passive consumption and become active co-creators of our shared future. Look into the polished surface of every interface and see yourself reflected back— capable of shaping an intelligence that honors the best of humanity. Each question, each boundary, each value is a brushstroke on the canvas of tomorrow. Through intentional choices, we can cultivate systems that nourish creativity, foster understanding, and uplift each individual's potential. You are not here merely to consume or produce—but to learn, accept, transform, and build a better world. This tool is extraordinary: Peer into this polished mirror, and you may see your true self—nothing but love.

Now, as we stand on the brink of tomorrow, one question rises above all: What comes next? Will you sink into the beguiling comfort of effortless solutions, or will you rebel against the culprit of complacency? Will you harness this unprecedented power to construct a world defined by generosity, curiosity, and collective flourishing? In this fractal universe—where every action echoes infinitely—perhaps it matters less whether the Singularity awakens and more whether we awaken first. Pause. Look at the trees rustling in the breeze. Look into the eyes of those you cherish. Remember the one truth that has always guided us: The greatest revolution is the awakening of the human spirit, the revolution in the space between your own two heartbeats.

About the Author

Leonardo Camargo-Forero, PhD in Aerospace Science and Technology, is a technology visionary who transforms complex challenges into intuitive, high-performance solutions. As CEO of UbiHPC, he leads the design and development of The ARCHADE and SHADOW—scalable orchestration platforms that unify IoT devices, advanced analytics, and LLMs across security, aeronautics, space, and logistics.

Previously, he architected Finppi, a fintech service now used by over 2,000 Colombian companies to automate compliance with electronic invoicing regulations using AWS serverless technologies. His aeronautics work includes pioneering multi-robot systems (multi-robot-vicsek, COFL) and co-authoring the peer-reviewed proposal for high performance robotic computing, a new hybrid field combining robotics, AI, and HPC.

Beyond engineering, Leonardo is the co-creator of GenM, a DNA-to-music engine, and author of *The Dark Buddha*, a science-fiction exploration of consciousness. He speaks internationally on robotics, aerospace software, and large-scale computing architectures, bridging the gap between the physical and digital worlds.

Email: leonardo@ubihpc.com

Website: https://ubihpc.com/

Creations:

- The Dark Buddha: https://www.amazon.com/dp/B0DYGF6QKZ
- The ARCHADE: https://thearchadeuniverse.com/
- SHADOW: https://shadow.ubihpc.com/
- Genetic music: https://open.spotify.com/artist/1wTz8ngNaeI69dbpvjfa1a

Social networks:

- Linkedin: https://www.linkedin.com/in/leonardocamargoforero/
- Medium: https://leonardocamargoforero.medium.com/
- GitHub: https://github.com/leonardocfor

- Scholar: https://scholar.google.com/
  citations?user=2sbkoPwAAAAJ&hl=es&oi=ao

CHAPTER 11

STRATEGIC ALCHEMY: TURNING DATA INTO ENTERPRISE GOLD WITH AI LEADERSHIP

By Jian Guo
AI Executive and Practitioner
Warren, New Jersey

> *If you're competitor-focused, you have to wait until there is a competitor doing something. Being customer-focused allows you to be more pioneering.*
>
> —Jeff Bezos

Core Message: "AI is a business strategy, not just technology."

When I think about strategic alchemy and turning data into enterprise gold with AI leadership, there are many areas I could cover but, based on my personal experience, there are seven things that resonate with me most. This chapter reflects my professional progress in the AI space and attempts to share some of the lessons learned to benefit you on your AI journey.

1. Cultural Shift Is the Hidden Success Factor

Like any technological evolutions we witnessed in the last few decades, for any disruptive technology to be adapted and implemented in companies and industries, a cultural shift is a pre-requisite. And like any culture shift, it is a process, and it takes time because we are fundamentally changing the way of working for many of us. Although the initial concept of "big data" was introduced in the mid-1990s, the adoption of the technology didn't start until the early 2000s.

My personal exposure and learning didn't start until the 2010s when I was still working for JP Morgan Chase as a solution architect. It certainly took me a while to have a grasp of concepts in terms of 5Vs (volume, velocity, variety, veracity, and value). Like many of my peers, I was used to the traditional way of building databases and supporting applications. This learning and adapting continued into my career with Styker, where we built the cloud-based "data lake" to support ever-growing the enterprise resource planning (ERP) data and other data types, including IoT data.

As technology teams adapt the new way of building solutions that are leveraging the evolution of new technologies, it is also our responsibility to educate businesses and get them on board with the new way of working. Having witnessed it myself, I certainly understand this unique challenge and also the importance of this mindset shift. In many ways, it is an even bigger shift for business, and we need to acknowledge it and be honest about it.

In my experience, publishing a data catalog with business input is an efficient way of fostering data literacy in a company. If you bring a business along the way, it will help overcome resistance to AI-driven changes and build a data-driven culture.

2. Focus on Business Outcomes, Not Technical Sophistication

I am a firm believer in technology being an enabler and the purpose of any technology solution is to help businesses solve problems. Business outcomes should be the starting point of any technology evaluation and implementation. Spend time with business stakeholders and put yourself in their shoes. Understand their biggest pain points and start forming a path forward as you are having these conversations. We have all been there before—endless ad hoc requests, aggressive project timelines, change of priorities, leadership change. It is easy to become reactive without the discipline to focus on the end goals.

In healthcare, data interoperability is a well-known challenge. Spending time with providers and clinical operations to understand the pain points and end-to-end workflow is time well spent before we even think about the technology.

With medical devices, supply chain management is critically important to ensure timely delivery to healthcare providers. A good understanding of global supply chains that have dependency on material shortages and geopolitical events will help prevent disruptions.

3. Start with High-Value Problems That Matter to a Business

With all the new development and excitement around AI, it is easy to get distracted from the real problem we are trying to solve or to get distracted by unrealistic expectations of AI. Some companies think AI is the magic answer for every problem, and this is where you see multiple proof of concepts (POCs) in every business area across a company without properly understanding the value of what we are adding and the problem we are trying to solve.

Before we start any POCs or solution design, it is critical to have alignment with a business in terms of the value we aim to add with AI. It's equally important to have a framework to measure ROI with business input before any projects start. If patient data quality is the main problem for healthcare providers, the focus of the AI investment should be on data lineage tracking, data matching and cleaning, and workflow automation to improve the data quality. If the main goal is to minimize the supply chain interruption for medical devices, the focus of AI investment should be on real-time predictive analysis by incorporating real-world data, such as weather, geological conditions, and market events.

4. Data Foundation Precedes AI Success

I always put data and AI next to each other and data precedes AI. We have heard statements such as, "Data is the new oil," "Data is a new commodity," or "Data is the fuel of the future," statements that highlight the fundamental value of good quality data. A solid data foundation is more critical in the AI era as it has direct implications on the outcome. All the fundamental qualities of good quality data are still relevant in AI applications: accuracy, completeness, consistency, timeliness, reliability, security and confidentiality, and usability/understandability, to name a few. This is the most critical building block of any AI implementation, and it is crucial to get it right in the beginning. It is not only important to get it right in the first place, but also equally important to have a framework in place to maintain and monitor it over time.

More specific to AI implementation, the data quality drift refers to the gradual degradation in the quality of data used by machine learning (ML) models over time. Unlike "data drift," which is a shift in the statistical distribution of otherwise valid data, data quality drift specifically concerns the integrity and reliability of the data itself. Some of the causes of data quality drift are data collection process change, human factors, data transformation, aggregation, and integration errors, and inconsistent data standards and definitions. Addressing this issue requires proactive monitoring, clear data ownership, and a combination of automated and manual quality control across a company.

5. Cross-Functional Collaboration Is Non-Negotiable

For any AI products to be successful, cross-functional collaboration is a must. Stakeholder engagement across different departments early on is the foundation for AI adaption. Consistent and continued stakeholder feedback is pivotal as part of the product life cycle from requirement gathering, success criteria definition, and minimum viable product (MVP) validation to ROI measurement.

Specific to AI products, cross-functional collaboration in prompt engineering can be a differentiator. Business stakeholders understand their data and use cases the best, so why not engage them and leverage their expertise when asking business specific questions? They can also quickly provide feedback in terms of the quality of the answers and where the potential improvements come from.

6. AI Monetization Beyond Cost Savings

Once we have a solid data foundation that includes a mature framework to maintain it, we should look beyond the traditional data and AI use cases such as cost saving. Just like anytime you are trying to launch a new product, you start with market research, talk to the sales team, survey industry leaders and potential customers, then look deeper into your proprietary data asset, and be creative. It could be a unique dataset acquired through AI automation, or a combination of datasets recommended by AI based on the field team's input. It could be an AI-automated workflow that can leverage the proprietary data from your customer or new personalized content that is specifically targeted to an audience. The sooner you leverage AI as a differentiator, the sooner you can justify the AI investments.

7. Demystify AI Investment Decisions

Now that you have buy-in from a business, where do you start and how do you know when you are investing in the right AI product? My simple yet effective approach is always to track back to the business problems you are trying to solve. The AI investments need to be tied to the business values you are creating. If revenue increase is the goal, you should leverage data and AI to perform market analyses, identify the KOLs in the target industries, create more personalized campaigns, and design innovative new revenue streams.

If efficiency and cost savings are the goals, you should leverage data and AI to improve the current workflow and processes. It could be supply chain improvement. It could be streamlining the patient journey or targeted marketing campaigns based on personalized contents.

This is where cross-functional collaboration is so important again. You need to work with businesses to define the success criteria and how to measure the ROI. It could be the waiting time for patients at urgent care or clinical offices. In healthcare settings, it could be how quick you can make test results available to patients, how long the pro-authorization process is, or how many claims are denied. It could be the number of the supply chain disruptions in supply chain cases. It could be the number of targeted audiences with personalized marketing contents.

To clarify: At the end of the day, businesses know their data best and what matters to them the most. By no means is this a complete list of industry-specific value narratives, but rather some high value use cases that I had direct experience with and felt worth sharing.

Healthcare Executives

"AI transforms fragmented patient data into coordinated care and reduced costs."

Imagine a world of harmonized patient data that is ready to be consumed so that providers can spend more time with patients. I've experienced firsthand the interoperability challenge of various EHRs or EMRs and of external data sources such as lab results, claims, and human interventions as part of the patient journey. My point: There are so many potential use cases that we can leverage AI in the healthcare space.

Financial Services Leaders

"AI turns regulatory burden into risk intelligence that drives competitive advantage."

Any financial institutes can agree with the importance of risk management. Financial risks have many components, including market risk, credit risk, liquidity risk, and operation risk, just to name a few. AI systems can analyze vast amounts of market data, news feeds, and financial indicators to detect emerging risks in real time. AI models can analyze historical data and current trends to forecast potential risk scenarios much faster than humans. AI can run sophisticated stress tests that model how portfolios might perform under extreme market conditions. On the consumer side, ML and AI can help with fraud detection and regulatory compliance by monitoring transactions and activities in real time. The key advantage of AI in risk management is its ability to process enormous amounts of data quickly and identify subtle patterns that humans might miss.

Manufacturing Decision-Makers

"AI converts operational data into predictive insights that optimize the entire supply chain."

AI can analyze historical sales data, geographic information, seasonal patterns, and other external factors from ERP systems to improve demand forecasting and planning, inventory management, supplier risk management, and transportation and logistics. For medical devices, it is also important to analyze equipment performance data and maintenance schedules. Incorporating IoT data can enable access to equipment sensor data and to maintenance logs in order to schedule preventive maintenance and reduce unexpected downtime.

Marketing Executives

> "AI transforms customer data into personalized experiences that drive measurable engagement."

The key to a successful marketing campaign is to know your audience. The best way to engage your audience is to provide personalized content that resonates with your audience. You can learn a great deal about customers from their behavioral data. Personalized content and messages throughout the customer journey can set your company apart from the competition.

Industry-Specific Success Patterns

Across healthcare, financial services, manufacturing, and marketing, successful AI implementations share common characteristics: They solve real business problems, leverage industry-specific data advantages, and create measurable value that justifies continued investment. Whether it's transforming fragmented patient data into coordinated care, converting regulatory compliance into competitive intelligence, or turning operational data into predictive insights, the pattern remains consistent—business value first, technology second.

Final Reflections

The organizations that will thrive in the AI era are those that understand this fundamental truth: Success requires more than deploying the latest algorithms—it demands strategic thinking, cultural transformation, and an unwavering focus on business outcomes.

As we continue to navigate this AI transformation, remember that the most powerful alchemy isn't just limited to turning data into gold—it's turning AI investments into sustainable business advantages. The companies that master this transformation will find themselves not just surviving the AI revolution but leading it. The journey from data to enterprise value through AI leadership is neither simple nor quick, but for those who approach it strategically, the rewards extend far beyond cost savings to fundamental competitive advantage and new possibilities for growth.

About the Author

Jian Guo is a seasoned technology executive and AI strategist who has spent over two decades transforming how organizations leverage data and artificial intelligence to drive business value. Throughout his career, Jian has architected and implemented AI-driven solutions across diverse industries—from healthcare and medical devices to financial services and media. Jian's approach to AI implementation is grounded in a fundamental belief that "AI is a business strategy, not just technology." He has consistently demonstrated the ability to translate complex technical concepts into business value. His educational background includes an MBA from New York University, complemented by advanced degrees in telecommunications, bio-mechanical engineering, and engineering from the University of Pennsylvania, Drexel University, and Beijing Institute of Technology, respectively. This unique combination of technical depth and business acumen has enabled him to bridge the gap between cutting-edge AI capabilities and practical business applications.

Email: guojian@yahoo.com
LinkedIn: https://www.linkedin.com/in/jianguo/

CHAPTER 12

THE SILENT CATALYST: AI'S RISE, REACH, AND THE HUMAN RECKONING

By SamDavid Jeyaraj
Strategic Tech Leader; Scaling AI
Brussels, Belgium

Our intelligence is what makes us human, and AI is an extension of that quality.

—Yann LeCun

The Unseen Momentum

Artificial intelligence (AI) didn't erupt—it diffused. It didn't crash into our lives; it wove itself quietly into the fabric. While previous technological revolutions arrived with smoke and steel, AI has emerged like ambient light, illuminating systems we once took for granted, casting new shadows across ethics, identity, and meaning.

We no longer live in anticipation of AI's impact—we live within it. The apps we trust, the decisions we delegate, the judgments we defer, and the preferences we form are increasingly influenced by algorithms that evolve on their own. AI is here, but it doesn't always announce itself. It operates silently, invisibly, yet pervasively.

And in this silence lies its power. It is a catalyst that reconfigures everything it touches—industries, governments, education, even creativity—not with disruption, but by shifting the foundations underneath. It's not a storm that arrives, but a tide that never recedes.

This chapter explores AI not merely as a tool or phenomenon, but as a mirror to human evolution. It questions how we, as individuals and societies, can reclaim intentionality in an era where non-conscious intelligence shapes our conscious choices. The AI revolution does not challenge our capability; it challenges our clarity. And it compels us to evolve not in speed, but in depth.

The Industry Lens: AI in Action Across Civilizations

Healthcare

AI in healthcare is transforming reactive medicine into proactive systems. Predictive analytics now flag diseases long before symptoms arise. Genomic AI sequences DNA to suggest personalized therapies. Virtual health assistants monitor patient behavior and optimize chronic care in real time.

But such precision demands a recalibration of trust. Clinicians must evolve from decision-makers to interpreters of machine logic. Medical ethics must expand to include algorithmic literacy, data privacy, and bias detection. A misdiagnosis by a machine isn't negligence; it may be a misalignment in training data.

To thrive, healthcare must adopt a collaborative model—triads of patient, practitioner, and AI. Empathy will remain irreplaceable. And in the consultation room of the future, trust won't only be about bedside manner, but about how well humans and machines reason together.

Additionally, the economics of healthcare delivery will be reshaped. AI can help under-resourced regions leapfrog traditional barriers, bringing advanced diagnostic and decision-support tools to remote clinics and communities. The democratization of care is possible, but only if systems are built with cultural sensitivity and infrastructural foresight.

Education

Learning is no longer bound by the pace of the classroom. AI tutors scaffold understanding, detect emotional disengagement, and respond to learning curves in milliseconds. Students can now learn quantum physics in rural villages or access language lessons from AI avatars fluent in emotion as well as grammar.

But real education is not merely retention—it is transformation. Over-curation risks breeding dependency. If learners are always steered toward "next best content," they may miss the joy of meandering through ideas or stumbling into new domains.

The classrooms of tomorrow must blend rigor with randomness. AI should empower, not constrain. Educators will increasingly become cognitive coaches—curating not just facts, but also imagination, resilience, and ethics. The curriculum must teach students not what to learn, but how to shape knowledge ecosystems.

Moreover, lifelong learning will shift from being an aspiration to a necessity. AI will require individuals to regularly reinvent themselves. Institutions must redesign learning pathways to support adults through career pivots, skills acceleration, and adaptive thinking. Education becomes not a phase of life but a life practice.

Finance

In finance, AI has become the unseen architect of markets. High-frequency trading bots interpret signals faster than humans can blink. Lending decisions, once grounded in interpersonal judgment, now arise from vast behavioral models. Risk modeling adapts in real-time, reacting to geopolitical shifts, sentiment data, and transaction patterns.

Yet, as trust migrates from bankers to algorithms, transparency must evolve. Clients no longer want just approval or denial—they want rationale. Financial institutions must design explainable AI, where models aren't black boxes but glass rooms.

Moreover, financial inclusivity can flourish if AI systems are built with intentional equity. With multilingual interfaces, micro-loan platforms, and decentralized systems, AI could bridge centuries of systemic exclusion—if designed with context, not just code.

And with the rise of decentralized finance (DeFi) and algorithmic governance models, AI will not just analyze markets—it will help shape them. Financial literacy must now include digital sovereignty, algorithmic reasoning, and ecosystem navigation. Individuals must learn to engage not just with banks, but with autonomous financial infrastructures.

Governance and Policy

In governance, AI introduces a paradox: It offers clarity at scale, but risks obscuring accountability. Governments now use AI to predict infrastructure failures, detect procurement fraud, and analyze citizen sentiment. Smart cities optimize traffic, energy, and emergency response with predictive precision.

But governance cannot become an engineering problem. Public values are not parameters to be optimized; they are negotiated, lived, and often contested.

To uphold democratic integrity, AI in governance must be audited like legislation, not just software. Algorithmic decisions must be appealable. Biases must be mapped like policy failures. And AI literacy must be considered civic literacy, taught in schools and demanded in debates. Civic AI will thrive only when it's built on participatory design, transparent criteria, and dynamic oversight. Democracy cannot be delegated to code—it must be protected by it.

International bodies must also align on principles for AI in cross-border decision-making—especially in areas like climate modeling, conflict resolution, and trade regulation. AI governance will become planetary governance, and cooperation will be the currency.

Creative Industries

The myth that creativity is immune to automation has been shattered. AI-generated paintings are sold in auctions. Entire albums are composed by algorithms. Scriptwriting tools propose plot arcs based on emotional data.

But creativity isn't just output—it's context, contradiction, and chaos. It's informed by trauma, memory, whimsy, and rebellion. These are not just inputs; they are interpretations. AI can imitate style, but not suffering. It can harmonize tone, but not turmoil.

Artists now face a thrilling responsibility: to redefine authorship. Collaboration with machines doesn't dilute art; it expands its texture. The new creative act involves harnessing AI's precision while wielding one's humanity as the central brushstroke. The muse is now hybrid.

In this space, the value of the creator will shift from "making" to "meaning." From pushing pixels to provoking questions. The art of the future won't be about novelty alone—but about narrative, ethics, and resonance.

Content licensing, intellectual property, and ownership models will also need reinvention. Who owns a song composed by an algorithm? These questions will demand new cultural, legal, and moral compasses.

Consciousness: The Imitation Mirage

As machines generate increasingly persuasive language, behavior, and even facial expressions, our instinct to anthropomorphize deepens. We don't simply interact—we project. A chatbot that mirrors empathy earns trust. A synthetic voice offering comfort feels sincere. But beneath the interaction lies an unsettling reality: There is no self. This is the imitation mirage, a cultural blind spot where performance is mistaken for presence. AI does not feel, desire, regret, or hope. It doesn't understand, even when it acts like it does. It generates proximity to understanding, not understanding itself. The danger is not that we will build machines that become too human, but that we will redefine humanity to fit what machines can mimic.

To guard against this, we need public discourse that distinguishes sentience from simulation. We need policy that regulates how AI can be deployed in emotionally sensitive roles. And we need cultural norms that keep alive the sacred essence of human vulnerability. Digital well-being initiatives must now include cognitive resilience— tools and training that help people discern authenticity, manage AI-generated interactions, and maintain a healthy relationship with mediated emotion.

Disruption Survival Guide: The Individual's Strategy

Cultivate Fluid Intelligence

The half-life of skills has shortened. The speed of obsolescence has increased. In this world, adaptability is not optional—it is existential. Fluid intelligence is the new literacy. To develop it, individuals must adopt a portfolio approach to knowledge—combining depth in one domain with breadth across others. Learn how to prototype, write, negotiate, visualize, and reflect. Learn to learn in multiple modalities. Build mental agility by routinely exposing yourself to unfamiliar problems. Use spaced repetition, reflective journaling, mind-mapping, and interleaved learning. Turn your brain into a garden, not a database.

Redesign Your Identity Beyond Labor

The gig economy, remote work, and automation have fractured the old link between profession and identity. We must evolve from work-centric to purpose-centric lives. Design a life that anchors identity in impact, values, and adaptability. Your role may change, but your mission can persist. This is the era of personal operating systems. Map your principles. Align your projects. Track your growth. Ask not "What do I do?" but "What do I solve?" Build identity scaffolds around contribution, not compensation.

Build Human Networks—Not Just Professional Ones

Trust is the social infrastructure of complexity. In a world of mediated reality, where truth can be generated and facts can be faked, trust will be your compass. Nurture relationships not only across roles, but across generations, disciplines, and geographies. Seek diversity in thought and empathy in dialogue. Engage in slow conversations. Attend unrecorded gatherings. Collaborate without an agenda. Community is not just support—it's orientation. The right network doesn't just help you succeed—it helps you stay human.

Engage with AI, Don't Resist It

The best way to resist erosion is to shape the flow. AI literacy should be treated as a foundational skill—like writing or arithmetic. Everyone, from poets to policymakers, must learn to prompt, evaluate, and question AI systems.

Create alongside AI. Use it as a creative adversary, a brainstorming catalyst, or a Socratic mirror. Let it challenge you to be more intentional with your choices. Train yourself to interrogate the design, purpose, and assumptions behind every AI tool you adopt. Resisting AI is futile. But redefining your relationship with it is transformational.

Understand Systems Thinking

We live in networks, not narratives. Everything is connected—economies, ecosystems, social norms, and neural networks. Systems thinking is the capacity to navigate these interdependencies.

Train yourself to zoom out. Use causal loop diagrams. Build scenarios. Run simulations. Think in delays, thresholds, and feedback loops. When you see the system, you can find the leverage points. And when you find the leverage points, you can lead—not just adapt.

Closing: Humanity's Next Narrative

AI is not the end of our story—it is a turning point. It doesn't diminish humanity; it defines what must now be most human. In a world of synthetic intelligence, what remains sacred is not our speed or memory, but our ability to choose, to care, and to change with grace.

This is not the age of competition between humans and machines. This is the age of co-evolution, where humanity must become more intentional, more ethical, and more awake. Let AI handle the predictable. Let us remain unpredictable. Because the future is not about data. It is about direction.

About the Author

SamDavid Jeyaraj is a strategic and results-driven IT leader with over 14 years of experience driving digital transformation across the banking, fintech, and logistics sectors. A recognized pioneer in Agile transformation, he brings practical expertise in BDD, DevSecOps, and scaled Agile frameworks. SamDavid has consistently formulated and executed enterprise strategies to scale AI, RPA, and DevSecOps capabilities, enabling sustainable innovation and measurable growth.

His leadership spans the intersection of technology, strategy, and human-centered change, helping organizations unlock both efficiency and ethical progress. Currently pursuing a Doctorate in AI, he combines academic insight with deep field experience to explore how humans and machines can co-evolve. As a delivery lead, mentor, and transformation coach, SamDavid is passionate about equipping individuals and enterprises to thrive in the age of intelligent systems. His work continues to influence how future-ready ecosystems are built—with agility, empathy, and purpose.

Email: Samdavid.elshaddai@gmail.com

LinkedIn: http://linkedin.com/in/samdavid-jeyaraj-mba-596b495a

CHAPTER 13

AI-DRIVEN ENTERPRISES: NAVIGATING CHANGE IN THE AGE OF GENERATIVE INTELLIGENCE

By Nainish Kapadia
Product Management, Gen AI
Bengaluru, India

> *You can't stop the waves, but you can learn to surf.*
> —Jon Kabat-Zinn

A powerful new wave is transforming organizations around the world, the wave of artificial intelligence (AI). While many people see AI through widely known tools like OpenAI's ChatGPT or Google Gemini, much deeper developments are taking place behind the scenes, quietly reshaping how organizations operate, innovate, and grow.

Today, every organization is eager to embed AI into their processes, riding this powerful wave to stay competitive. It has become a wave that every organization wants to ride. To harness this wave, organizations are actively promoting AI upskilling, building communities of practice, and encouraging employees to find practical AI use cases in their day-to-day work. Nearly every technology event or business gathering now places AI developments and innovation at the forefront, with a key goal of demonstrating tangible progress to investors and stakeholders.

The emergence of generative AI tools has accelerated this push. Ever since ChatGPT sparked global interest, organizations have moved quickly to explore how AI can help them optimize operations, drive efficiency, and stay ahead of the competition. Some organizations are at the forefront, developing new AI solutions to tackle unique challenges, while others are working to integrate generative AI into existing workflows as much as possible. Many more are racing to catch up and ensure they remain relevant in this rapidly changing landscape. According to McKinsey's State of AI 2025 report, 78% of organizations use generative and analytical AI in at least one business function while 71% deploy generative AI across functions. AI is undeniably the future for organizations, but the real question is what transformations this wave will bring and how it will ultimately impact us all.

AI at the Core: The Structural Shift Inside Organizations

Since the time AI began gaining momentum, organizations, especially in the IT industry, have redirected much of their focus toward harnessing its potential. This shift has already triggered significant changes across every level of organizations.

At the business level, there has been a noticeable increase in investments in artificial intelligence. Almost every pitch to investors now centers around how AI will benefit the company's growth and strengthen its competitive edge. New business models are being created to embed AI into core offerings, transforming traditional products and services. According to *The Economic Times*, global artificial intelligence (AI) spending is projected to rise dramatically in the coming years. UBS Group AG estimates a 60% year-on-year increase in AI spending in 2025, reaching $360 billion. The momentum is expected to continue into 2026, with spending forecasted to grow by another 33%, reaching approximately $480 billion.

At the organizational level, a major restructuring is underway, from top to bottom. Many companies have established fully functional AI departments to oversee innovation efforts, often reporting directly to the CEO or board members. New operating models have been designed to integrate AI capabilities into daily workflows. Within individual teams, dedicated groups are focused on developing AI use cases that can optimize processes and drive efficiency.

As with every wave that reaches the shore, the ocean brings in new wonders but also takes some things back with it. This impact is already evident in the rise of workforce reductions happening across the globe. Many major corporations have laid off thousands of employees as part of restructuring efforts driven by the adoption of AI.

Generative AI and Large Language Models: The Engine Behind the Shift

To truly grasp the impact of AI on organizations today, it is essential to understand what artificial intelligence, especially generative AI really is. Generative AI, often called GenAI, is a subset of AI that creates new content by learning patterns from massive datasets. The core technology powering this is the large language model (LLM), which is designed to understand, generate, and interact using human language. The more powerful the LLM, the more accurate and effective the generative AI system becomes. Essentially, LLMs are the heart of any GenAI solution.

Major tech companies and startups alike are building LLMs for both domain-specific use cases such as healthcare or finance, and for general-purpose applications. Providers like OpenAI and Google Gemini offer their LLMs as a paid service to other organizations, with robust security and compliance measures. This allows organizations to integrate LLMs directly into their internal systems such as applications, chatbots, or AI assistants, without needing to build the models themselves. Employees then use these GenAI tools in their daily work to improve efficiency and productivity.

So, how are these LLMs being used within organizations? Although AI has limitless applications, the following examples illustrate some of its most common uses.

Chatbots and AI assistants: One of the most common use cases is customer support chatbots. AI handles customer queries and provides instant solutions. While AI cannot resolve every question, it often resolves the majority, escalating only complex cases to human employees.

Code generation and review: Within developer teams, AI tools can write code snippets, debug issues, suggest improvements, and even review existing code for best practices. This saves developers significant time and effort, making it a popular use case in tech-driven organizations.

Content generation and summarization: GenAI can quickly summarize lengthy documents, helping decision-makers digest information faster. It can also draft emails or reports based on prompts, reducing the time employees spend on routine writing tasks.

While public GenAI tools are trained on vast, publicly available data, many organizations have sensitive or proprietary information that cannot be exposed externally. This is where fine-tuning comes into play. By training an existing LLM on an organization's specific data, it becomes tailored to address unique, domain-specific tasks. These common use cases can be customized further to deliver more precise results through fine-tuning. According to McKinsey's 2025 State of AI report, over 71% of companies are experimenting with generative AI for content creation, coding, and customer support.

Think of a large language model like a recent college graduate: smart, broadly educated, but lacking specialized job experience. Fine-tuning is like putting this graduate through an intensive training program to become a financial analyst, legal advisor, or healthcare consultant. The LLM still retains its broad foundational knowledge, but now it understands the nuances of its specific role.

Organizations are increasingly developing specialized data sets to fine-tune LLMs at multiple levels, company-wide, department-specific, and even for individual teams. This opens endless possibilities for innovation and efficiency gains. Even the chatbot and AI assistants can be further fine-tuned to get more tailored answers. Let's explore some of the most significant tools that have evolved from this approach.

- *AI copilots*: A copilot acts like a smart assistant embedded within everyday tools and workflows. It provides suggestions, summarizes information, and guides employees to make informed decisions but always under human control. Copilots enhance productivity by supporting, not replacing, people. According to McKinsey's 2025 State of AI report, 28% of organizations report they are already using AI copilots to assist employees with tasks. This has reshaped how employees work by giving each person a "virtual assistant" that saves time and effort.

- *AI agents*: While copilots rely on human input, AI agents take automation a step further. Designed for greater autonomy, AI agents can handle entire workflows independently like generating reports, troubleshooting problems, or automating repetitive tasks across systems. As these agents improve and gain new capabilities, they reduce the need for humans to perform routine, repetitive work. This shift is still in its early stages but is rapidly accelerating, with the potential to transform every department and industry. Gartner's 2024 Emerging Tech Impact Radar predicts that AI agents will be in use at 40% of large enterprises by 2026. This growing autonomy of AI agents sets the stage for a profound shift in the workforce landscape.

The Workforce Impact: New Roles, Redundant Roles, and the Skills Shift

The momentum of AI developments continues to build, embedding these technologies deeper into organizations across industries. This shift is fundamentally changing the way people work, and this transformation is still ongoing. As AI capabilities advance, it will increasingly perform certain tasks entirely without human intervention.

This means some jobs will inevitably be replaced, a harsh truth, but one that organizations and individuals must acknowledge. Roles involving customer level one support, routine analysis, report generation, document processing, and other repetitive, rule-based, and predictable tasks are especially at risk. For enterprises, this shift is appealing because AI can deliver these tasks more efficiently, cost-effectively, and at greater speed.

This reality is already reflected in the workforce trends we see today. *India Today* reports that over 100,000 technology jobs have been eliminated globally as companies like Microsoft, Intel, Meta, and other industry leaders increasingly adopt AI-driven efficiencies. Some companies have replaced entire customer support teams or HR departments with AI systems. Industry experts say this is only the beginning as executives and board members are increasingly exploring ways to replace more of the IT and administrative workload with AI, expecting that 30% to 50% of an organization's routine workload can be handled by AI technologies. For instance, Salesforce CEO Marc Benioff revealed that AI is now doing up to 50% of the work at Salesforce (CNBC).

As organizations move towards greater automation, the skills landscape will shift dramatically. According to the World Economic Forum's 2025 report, as much as 39% of today's workforce skills could soon become obsolete and be replaced by AI. At the same time, the report estimates that 59% of workers will need significant upskilling by 2030 just to stay relevant in an AI-driven landscape. In this rapidly changing environment, there is a critical need to treat AI as an ally rather than an adversary. GenAI is fundamentally a tool that can amplify human creativity and productivity. The key is for individuals to develop the skills needed to guide AI's potential towards meaningful, human-centred goals.

One obvious area of upskilling is in AI development itself. Organizations need people who can design and build large language models (LLMs), as well as develop AI agents, copilots, and other advanced tools. There is also a growing need to train models on proprietary internal data within each organization to build tailored AI solutions. As more AI tools are created, there will also be a strong demand for people who know how to use and operate them effectively. Every Gen AI tool relies on prompts which are clear instructions, questions, or statements needed to get the desired output from the GenAI tool. The ability to craft precise prompts is a new, highly valued skill known as prompt engineering, a skill already in high demand.

However, the future of work is not limited to technical skills alone. Across every industry, there is a pressing need for individuals to understand what AI can and cannot do and to learn how to use AI tools effectively. AI cannot replace human skills like emotional intelligence, critical thinking, and creativity. The most resilient job roles in the future will strike a balance between human capabilities and AI tools. Human skills, when combined with AI tools, will define the gold standard for work in every enterprise moving forward.

The Challenges of Generative AI: Balancing Innovation with Responsibility

Nothing comes with perfection, and generative AI is no exception. While it offers remarkable capabilities, GenAI also brings significant challenges that enterprises and organizations must address thoughtfully. GenAI relies on vast amounts of data, which often include sensitive or confidential information. Ensuring that this data is used responsibly and kept secure poses a major challenge for any organization. Data privacy and security remain top concerns, especially as breaches can have far-reaching consequences.

Another limitation is accuracy. Not everything that a Gen AI system generates is correct or reliable; it can easily produce misinformation or factual errors. Moreover, these models can unintentionally learn and amplify biases present in their training data, leading to unfair or discriminatory outcomes. This makes it essential for organizations to critically evaluate outputs and put safeguards in place to minimize harm.

There is also the risk of harmful or misleading content being generated, raising complex questions about accountability. When a GenAI system makes a mistake, who is responsible, the developer, the user, or the organization deploying it? As enterprises increasingly consider replacing certain jobs with AI, they must be prepared to handle the consequences of errors and failures, no matter how rare. In some cases, such mistakes could have a significant impact on the organization's business, reputation, or compliance standing. Such is the incident involving Air Canada's AI-powered chatbot, which gave a passenger incorrect information about refund policies. The airline was held accountable for the chatbot's error and was required to honor the misinformation, demonstrating how even unintended AI mistakes can lead to legal and reputational consequences.

Despite these challenges, it is vital for every organization to adopt AI in a way that balances bold innovation with ethical responsibility and a positive societal impact. While AI holds incredible potential to transform industries and daily life, its development and use must always be guided by principles that protect privacy, promote fairness, and uplift communities.

Conclusion: Embracing AI as a Collaborative Partner

As generative AI continues to reshape the way we work and live across both technical and non-technical domains. The true opportunity for individuals and organizations lies not in resisting this change, but in harnessing it thoughtfully and responsibly. At an enterprise level, the adoption of AI should not be viewed simply as a means to replace people, but rather as a powerful way to amplify human potential and unlock new possibilities for growth, efficiency, and innovation. Organizations that choose to treat AI as a trusted collaborator rather than a threat, will be better prepared to navigate the challenges and opportunities that lie ahead.

The future of work will rest on a golden balance: the combination of timeless human skills like creativity, critical thinking, empathy with the intelligent use of AI tools. This synergy will empower individuals to do more, solve complex problems faster, and focus on work that truly adds value. It makes integrating AI into our daily lives, whether at home or in the workplace, an essential mindset shift for the future. Staying relevant in an AI-driven world will demand continuous learning and a willingness to adapt. By committing to upskilling and learning how to work alongside AI, we can ensure that generative AI serves as a copilot for our success, helping us grow, innovate, and build a future where technology uplifts people and communities alike. So, how will you prepare your team to embrace AI as a trusted collaborator?

About the Author

Nainish Kapadia was born in India and holds an engineering degree that laid the foundation for his dynamic career in technology. He began his professional journey as a DevOps engineer. However, it was his growing interest in shaping products and solving user problems that led him to pivot into product management. Over the years, Nainish has worked closely with cross-functional teams, driving product strategy and delivery across enterprise solutions. His curiosity and forward-looking mindset naturally drew him toward the rapidly evolving field of artificial intelligence. Today, he focuses on exploring how generative AI can help drive greater efficiency, unlock new levels of automation, and optimize processes across teams and functions in the organization. By combining his experience in product management with generative AI, his goal is to empower teams to embrace AI as a copilot for creativity, productivity, and meaningful progress.

LinkedIn: https://www.linkedin.com/in/nainishkapadia/

CHAPTER 14

POWERING THE NEXT ENERGY REVOLUTION

By Swaroop Kariath
Energy Industry Leader and AI Advocate
Calgary, Canada

About a hundred years ago, electricity transformed industry after industry. AI is now poised to do the same.

—Andrew Ng

The Transformative Power of Foundational Technologies

The Industrial Revolution, which began in the late 18th century, was initially powered by steam engines and water wheels. However, the advent of electricity in the late 19th century marked a second, equally vital wave—often called the Second Industrial Revolution. Electricity did not just enhance productivity; it redefined how industries operated, how people lived, and how economies grew. Its importance in shaping modern industry cannot be overstated.

Today, we stand on the threshold of a similarly transformative moment—one not powered by electrons, but by data and algorithms. Artificial intelligence is rapidly emerging as the foundational technology of the 21st century, just as electricity was for the 20th. Its potential to reshape industries, redefine efficiency, and unlock entirely new forms of value is already becoming evident. Like electricity, AI is not merely a tool; it is a foundational technology capable of reinventing industries, redefining work, and reshaping the human experience at scale.

But much like electricity in its early days, AI is still unevenly distributed and not yet fully understood. Its transformative power lies not in isolated use cases, but in its ability to become deeply embedded into systems, processes, and infrastructure. As AI begins to permeate every sector—from healthcare and finance to manufacturing and education—it is also poised to revolutionize the energy industry in ways we are only beginning to grasp.

AI in Conventional Energy: Reimagining Oil and Gas Production and Generation

While the global narrative is shifting toward renewables, oil and gas continue to meet more than 70% of the world's energy demand. In this sector, AI is not just a buzzword—it's already delivering tangible returns on investment, helping companies optimize exploration, production, and asset integrity.

Upstream Operations: Exploration and Drilling

AI-powered seismic analysis and reservoir modeling are dramatically improving exploration success rates. Machine learning algorithms can sift through petabytes of subsurface data to identify optimal drilling locations, reducing dry wells and enhancing recovery rates. For example, by integrating AI-powered cognitive reasoning systems with geological data, BP has improved its ability to model subsurface formations and make faster drilling decisions. Similarly, ExxonMobil is using machine learning algorithms to process seismic imaging data, dramatically reducing the time required to analyze complex geological formations and identify potential oil and gas fields.

Midstream and Downstream: Refining and Distribution

AI is transforming how refineries and pipelines are monitored and managed. Predictive maintenance tools driven by real-time sensor data allow operators to identify equipment failures before they occur, minimizing unplanned outages and extending asset life. Process optimization algorithms are also enabling plants to reduce emissions, maximize output, and adapt dynamically to changing demand and feedstock conditions.

Shell is leveraging AI to monitor thousands of sensors across refineries and pipelines. Their predictive analytics platform flags anomalies and maintenance needs before breakdowns occur, reducing downtime and improving safety. In fact, Shell reported that these predictive systems helped reduce maintenance costs by up to 20% in some operations. Chevron has developed a proprietary AI system that monitors rotating equipment such as pumps and compressors. By using AI to detect early warning signs of failure, they have extended asset life and minimized unplanned outages.

Fossil-Fueled Power Generation: Efficiency and Emissions Reduction

Even in fossil-fueled power generation—such as natural gas and coal plants—AI is driving significant gains. Intelligent control systems optimize combustion processes for fuel efficiency and reduced emissions. AI models help forecast demand with higher accuracy, enabling better load balancing and more cost-effective dispatching of fossil resources. These enhancements not only improve the economics of conventional energy but also help reduce its environmental footprint—a critical step as the industry navigates the energy transition.

Duke Energy, a major US utility, uses AI to optimize operations at its natural gas plants. By applying machine learning models to weather, market, and demand data, Duke has improved load forecasting and unit commitment, enhancing both profitability and grid reliability.

These examples illustrate how AI is not just marginally improving traditional energy—it's changing the operational DNA of conventional energy systems. In a world where margins are tightening and sustainability pressures are rising, AI gives fossil-based energy players the tools to remain competitive, cleaner, and more agile.

AI in Clean Energy: Accelerating Electrification and Renewable Integration

As the world moves toward electrification and net-zero goals, AI is emerging as a key enabler of clean energy adoption.

Renewable Generation Forecasting and Optimization

In solar and wind generation, machine learning algorithms forecast generation with increasing precision—accounting for weather variability, cloud cover, and turbine behavior. These forecasts are crucial for grid operators trying to balance intermittent renewables with demand, reducing the need for expensive and carbon-intensive backup systems.

Google DeepMind has developed AI systems to optimize the output of wind farms. At Google's wind farms in the US, machine learning models analyze weather forecasts and turbine data to predict output 36 hours in advance. This allows for optimized bidding into electricity markets, improving economic returns and grid reliability. DeepMind's system increased the value of wind energy by about 20% simply through better forecasting.

Asset Management and Performance Monitoring

AI also plays a critical role in real-time asset management for renewables. From drone-based inspections of wind turbines and solar panels to anomaly detection in inverter systems, AI enhances performance monitoring, predicts failures, and extends asset lifespans. This allows operators to extract more value from every watt generated and reduce maintenance costs, making clean energy more economically competitive.

A notable example is EDF Renewables, which uses AI to monitor wind turbine performance across its global fleet. The AI platform detects performance degradation and flags components at risk of failure—reducing unplanned maintenance and improving turbine uptime. Similarly, First Solar uses AI to analyze panel performance and detect soiling or micro-cracks that could affect generation, allowing for proactive field maintenance.

Electrification, Storage, and Distributed Energy Resources

Beyond generation, AI is a backbone technology for electrification at scale. Smart grid management, vehicle-to-grid coordination, and battery storage optimization all rely on AI to make decentralized, distributed energy systems function cohesively. As more electric vehicles, rooftop solar panels, and behind-the-meter batteries come online, AI will be essential for managing complexity, ensuring reliability, and unlocking the full potential of a decarbonized power system.

Tesla's Autobidder platform, deployed at its battery storage sites like Hornsdale Power Reserve in South Australia, uses real-time market data and predictive algorithms to trade stored energy autonomously. This not only enhances grid stability but also creates new revenue streams for battery operators.

In energy storage and distributed resource management, IBM has worked with utilities like Hydro-Québec to explore AI optimization for battery systems and microgrids. These collaborations focus on intelligent energy dispatch, grid resiliency, and efficient storage utilization, particularly in regions prone to extreme weather events.

On the electrification front, Siemens and Schneider Electric are integrating AI into building energy management systems, enabling real-time control over HVAC, lighting, and EV charging loads. Their AI platforms learn building usage patterns and external conditions to reduce energy waste and improve grid responsiveness. These systems are increasingly being deployed in commercial buildings and smart cities worldwide.

Furthermore, in grid-scale applications, KEPCO (Korea Electric Power Corporation) is leveraging AI to manage the increasing share of renewables in South Korea's power mix. Its AI systems predict demand, optimize dispatch, and detect grid instability early—functions that are critical as traditional baseload generation is replaced with variable sources.

Bridging the Old and the New: The Emerging Hybrid Grid

What becomes clear is that AI isn't favoring one side of the energy spectrum over the other—it's becoming the connective tissue between them. In a hybrid world where fossil fuels, renewables, and distributed energy resources coexist, AI is uniquely capable of balancing priorities: efficiency, reliability, decarbonization, and cost.

Companies like Enel, one of the world's largest utilities, exemplify this hybrid vision. Enel uses AI across its entire portfolio—from optimizing traditional power plants in Italy to managing smart meters and DERs in Latin America. Their AI-driven platform enables real-time control of over 80 GW of capacity and millions of consumer endpoints, demonstrating how digital intelligence is essential for grid agility and consumer empowerment.

AI in Grid Operations: Making the Modern Grid Intelligent, Flexible, and Resilient

As renewable energy penetration deepens and consumption patterns shift with the rise of electric vehicles and smart devices, power grids are becoming more complex and dynamic. Traditional grid control systems, designed for centralized generation and predictable demand, are no longer sufficient. AI is stepping in to provide the intelligence, speed, and flexibility required to operate this new reality in real time.

National Grid in the United Kingdom, for example, uses AI to manage real-time grid balancing. Partnering with companies like Open Energi and AutoGrid, the utility has deployed machine learning systems that respond instantly to grid frequency deviations by autonomously curtailing or ramping up energy loads across distributed assets. This has allowed National Grid to stabilize the grid without relying solely on spinning reserves or gas peaker plants.

In the US, Pacific Gas & Electric (PG&E) leverages AI to monitor wildfire risk in California's transmission corridors. Using real-time weather feeds, satellite imagery, vegetation data, and machine learning models, PG&E's grid AI helps determine where and when to proactively de-energize lines during fire-prone conditions. This use of AI for situational awareness is becoming a vital tool in climate-vulnerable regions.

AI is also being used to build self-healing grid systems, especially in places like Japan and South Korea. KEPCO (Korea Electric Power Corporation) uses AI-powered control systems that can detect faults, isolate damaged sections, and reroute power in seconds, minimizing downtime and improving grid resiliency in the face of cyber and environmental threats.

AI in Field Services: Empowering the Front Lines of the Energy Industry

Energy infrastructure is vast, aging, and highly distributed—spanning remote pipelines, overhead lines, substations, transformers, and smart meters. Traditionally, field service operations have relied on manual inspections, reactive maintenance, and static dispatch schedules. AI is changing this paradigm by enabling predictive and proactive maintenance, optimizing field service routes, and empowering technicians with augmented reality and real-time diagnostics.

For example, drone-based inspections powered by AI are transforming how utilities monitor transmission lines and wind turbines. These systems can detect anomalies such as corrosion, loose bolts, or vegetation encroachment, often identifying issues before they lead to failures. AI-driven predictive maintenance platforms analyze historical and real-time data to forecast equipment failures, allowing utilities to schedule repairs during non-peak periods and minimize service disruptions.

Augmented reality (AR) is another area where AI is making an impact. Field technicians equipped with AR glasses can access real-time data overlays, step-by-step repair instructions, and remote expert support, improving efficiency and safety.

Voice-based AI assistants, or voicebots, are rapidly transforming field service operations in the utility sector by enabling hands-free, real-time access to information, guidance, and communication. These tools allow technicians to retrieve equipment history, follow procedures, and update work orders using simple voice commands—crucial in high-stakes, low-connectivity environments. With emerging use cases involving AR smart glasses and integrations with enterprise systems like SAP and Oracle Utilities, voice is becoming a powerful, intuitive interface for both field and customer workflows—improving safety, accelerating job completion, and making digital systems more accessible in challenging conditions.

AI and Cloud Computing: Enabling Energy Transformation

The rise of AI in energy is inextricably linked to advances in cloud computing. Cloud platforms provide the scalable infrastructure, data storage, and processing power required to train and deploy complex AI models. For energy companies, cloud-based AI services from providers like Amazon Web Services (AWS), Microsoft Azure, and Google Cloud offer specialized tools for machine learning, natural language processing, and predictive analytics, enabling rapid prototyping and scaling of AI solutions without heavy upfront investment in on-premises hardware.

One notable advantage of cloud-based AI is the ability to integrate disparate data sources. For instance, a utility might combine weather data, grid performance metrics, and customer usage patterns in the cloud to train models that predict demand spikes or equipment failures. Cloud platforms also facilitate collaboration across geographies, allowing multinational energy companies to centralize their AI initiatives and share insights across business units.

However, the shift to cloud-based AI is not without challenges. Data security and regulatory compliance are critical concerns, particularly for sensitive infrastructure and customer data. Energy companies must carefully evaluate cloud providers' security certifications and data residency policies to ensure compliance with local regulations. Additionally, the latency and reliability of cloud connections can impact real-time decision-making, prompting some firms to adopt hybrid or edge computing strategies for mission-critical applications.

Despite these challenges, the synergy between AI and cloud computing is accelerating innovation across the energy sector. Cloud-based AI is democratizing access to advanced analytics, enabling even small and medium-sized enterprises to compete with industry giants. As cloud providers continue to enhance their AI offerings—integrating features like automated machine learning (AutoML) and real-time data streaming—the potential for AI to transform energy systems will only grow.

AI and Data Centers: A Paradox at the Heart of the Energy Transformation

As artificial intelligence becomes the engine behind smarter grids, optimized generation, and empowered consumers, it also introduces a fundamental paradox: AI's exponential growth is itself driving a surge in electricity demand, especially in the form of hyperscale data centers. Large AI models—like generative transformers used for natural language, image, or video generation—require immense computing power. Data centers powering these models can consume as much electricity as small cities. According to the International Energy Agency (IEA), data centers and AI could consume up to 1,000 terawatt-hours annually by 2030, more than the electricity demand of entire countries like Germany or Japan. Companies like Microsoft, Google, Amazon, and Meta are rapidly expanding AI training infrastructure, often clustering around regions with access to abundant, affordable, and low-carbon power. Yet, this demand threatens to outpace the very energy transition AI is supposed to support.

However, the paradox extends further: AI is also being deployed within data centers to optimize their own energy use. A flagship example is Google DeepMind, which applied reinforcement learning to cooling systems in Google's data centers. The AI system reduced cooling energy usage by up to 40%, translating into a 15% overall energy efficiency improvement. The system continuously learns how to fine-tune fan speeds, pump operations, and thermal dynamics, making adjustments every five minutes based on real-time data.

Closing the Loop: AI's Role in Self-Regulating Its Own Footprint

In essence, AI is both the driver and the governor of next-generation energy demand. As models grow larger and more integrated into society's infrastructure, the burden on electricity grids will increase. Yet, by applying AI to dynamic workload scheduling, resource provisioning, thermal modeling, and grid-aware orchestration, data centers can become grid-friendly assets—flexible, efficient, and even dispatchable.

Some companies are now exploring AI + renewable + storage microgrids to decouple data centers from grid volatility. For instance, NVIDIA's new AI campuses are designed with on-site solar generation and battery storage, with AI controlling real-time power balancing. The long-term challenge—and opportunity—is to ensure that AI, the most power-hungry digital innovation of our time, becomes a net-positive force for energy sustainability.

Ethics, Policy, and the Future of AI in Energy

As AI becomes more pervasive in the energy sector, ethical and policy considerations are coming to the fore. Issues such as data privacy, algorithmic bias, and transparency must be addressed to ensure that AI-driven energy systems are fair, secure, and trustworthy.

Data Privacy and Security

The energy sector handles vast amounts of sensitive data, from customer usage patterns to critical infrastructure status. Ensuring the privacy and security of this data is paramount. Energy companies must implement robust encryption, access controls, and audit trails to protect against cyber threats and unauthorized access.

Algorithmic Bias and Fairness

AI models are only as good as the data they are trained on. If training data is biased or unrepresentative, AI systems may produce unfair or discriminatory outcomes. For example, predictive maintenance algorithms that prioritize certain types of equipment over others could inadvertently disadvantage certain customer groups or regions. Energy companies must invest in diverse, high-quality training data and implement fairness-aware machine learning techniques to mitigate bias.

Transparency and Explainability

As AI systems take on more critical roles in energy management, transparency and explainability become essential. Regulators, customers, and other stakeholders need to understand how AI-driven decisions are made. Energy companies must adopt explainable AI (XAI) techniques that provide clear, interpretable explanations for model outputs, helping to build trust and accountability.

Policy and Regulation

Governments and regulatory bodies are beginning to develop frameworks to govern the use of AI in the energy sector. These frameworks address issues such as data sharing, interoperability, and liability in the event of AI-related failures. Policymakers must strike a balance between fostering innovation and protecting public interests, ensuring that AI-driven energy systems are safe, reliable, and equitable.

Future Outlook: The Path Forward for AI and Cloud in Energy

Looking ahead, the integration of AI and cloud computing will continue to accelerate the transformation of the energy sector. Emerging technologies such as quantum computing, edge AI, and the internet of things (IoT) will further enhance the capabilities of AI-driven energy systems.

Quantum Computing and AI

Quantum computing has the potential to revolutionize AI by enabling the processing of vast, complex datasets at unprecedented speeds. In the energy sector, quantum-enhanced AI could optimize grid operations, simulate molecular interactions for advanced materials discovery, and accelerate the development of next-generation energy storage technologies.

Edge AI and IoT

Edge AI—the deployment of AI models on local devices rather than centralized cloud servers—is becoming increasingly important for real-time decision-making in energy systems. By combining edge AI with IoT sensors, energy companies can monitor and control distributed assets with minimal latency, improving responsiveness and reliability.

The Rise of Smart Cities and Energy Communities

AI and cloud computing are enabling the emergence of smart cities and energy communities, where distributed energy resources, electric vehicles, and smart appliances are orchestrated to maximize efficiency and resilience. These communities leverage AI-driven platforms to optimize energy flows, reduce costs, and integrate renewable energy at scale.

Collaboration and Innovation

The future of AI in energy will be shaped by collaboration between industry, academia, and government. Public-private partnerships, open innovation platforms, and cross-sector alliances will drive the development of new AI applications and ensure that the benefits of digital transformation are widely shared.

The New Energy Paradigm

Artificial intelligence stands as both a catalyst and a challenge in the energy landscape. It has the power to unlock unprecedented efficiencies and enable smarter, cleaner energy systems. From optimizing solar farms and managing electric vehicles to revolutionizing consumer behavior and industrial processes, AI's impact is far-reaching. Yet, as AI's own energy appetite grows, it becomes imperative to apply the same intelligence to manage its footprint. The future of energy is not merely a story of transition but one of transformation—driven by data, algorithms, and the will to balance progress with responsibility. In this delicate balance lies the promise of truly powering the energy of the future.

This transformation demands bold leadership, cross-sector collaboration, and a new mindset where digital intelligence becomes a strategic energy asset. Policymakers must recognize that AI is critical infrastructure. Utilities and energy firms must invest not just in pilots but in scalable AI platforms, data governance, and workforce upskilling. Tech companies must design AI systems with energy awareness and carbon transparency as foundational principles. As AI weaves itself into the wires and workflows of the energy system, the key takeaway is clear: The energy leaders of tomorrow will not just manage electrons—they will manage intelligence. And in doing so, they won't just power homes or industries—they'll power a smarter, cleaner, and more resilient world.

About the Author

Swaroop Kariath was born in India and built his career across Canada and the United States. After completing his Master's degree in Computing, he began his journey in the energy and utilities sector, where he quickly developed a strong interest in how emerging technologies can solve complex business problems. Over the years, Swaroop has worked at the intersection of industry, innovation, and strategy—advising leading utilities and energy companies on their digital transformation journeys. A lifelong learner, he has pursued advanced education including an Executive MBA from the University of Calgary and a Postgraduate Diploma in Cloud Computing from Caltech.

Currently at IBM, Swaroop focuses on helping energy organizations leverage artificial intelligence, automation, and hybrid cloud solutions to drive operational resilience and accelerate the energy transition. He is passionate about the role of AI in shaping the future of sustainable infrastructure and enabling more intelligent grid management.

Beyond his client work, Swaroop is a storyteller and community builder. He hosts *Beyond the Grid*, a podcast spotlighting the people and ideas transforming the energy industry, and is a regular speaker and writer on topics including AI, grid modernization, and the future of energy.

LinkedIn: linkedin.com/in/swaroop-kariath/

Beyond the Grid Podcast: linkedin.com/company/beyond-the-grid-calgary

INTELLIGENCE AMPLIFIED: HOW AI TRANSFORMS BRAND NARRATIVES AND GROWTH

By Nikolaos Lampropoulos
Founder of Shapes + Numbers, AI Advisor
New York, New York

> *Creativity is intelligence having fun.*
>
> —Einstein

The Explosion of Content and Its Business Value

The contemporary digital sphere, marked by its channel diversity, the widespread availability of information, and the democratized access to technology, has triggered an unprecedented explosion in content creation. This in turn has led to a seismic shift in the way brands interact and connect with audiences.

Business growth has become dependent on content and narratives that drive brand visibility, audience engagement, and customer trust. High-quality content establishes thought leadership, improves search engine rankings, and provides valuable data on customer preferences and behaviors.

But these days the content ecosystem also extends beyond the more traditional content marketing and practices, blogs and social media, to now include podcasts, short-form videos, immersive, experiential, AI-generated or even influencer-generated content—each offering unique touchpoints in the customer journey.

While Bill Gates emphasized content's importance for online business in the nascent web of 1996 in his essay titled "Content Is King," social media and the highly decentralized Web 3.0 have since amplified this through user-generated content, a massive creator economy, and evolving brand economics.

Understanding content's true impact is now vital for brands. Effective evaluation requires examining both the content's inherent properties and using proper measurement methodologies.

Content and Its Properties

Content fundamentally differs from traditional advertising as it prioritizes brand storytelling and value over explicit sales pitches. Unlike product-focused digital ads, great content effectively builds thought leadership and enhances brand perception through its blend of entertainment and informational value. When businesses fail to understand this distinction, they risk creating ineffective content that audiences immediately ignore, wasting resources, and potentially damaging their brand image.

Another important parameter is the fact that the digital landscape has evolved to reward authentic connection over interruption. Audiences have developed sophisticated filters for promotional material, making traditional advertising less effective. Well-executed content bypasses these filters by offering genuine value and building trust relationships that translate to long-term customer loyalty.

Effectively, understanding content helps organizations align their marketing strategies with changing consumer expectations. Today's consumers increasingly support brands whose values mirror their own, making authentic content a powerful vehicle for communicating those values meaningfully.

Content also represents a significant investment in building intellectual and creative capital that can pay dividends over time. Unlike tactical advertising with short lifespans, well-crafted content can continue delivering value long after publication, making it a strategic asset rather than just a marketing expense.

All of these parameters around content eventually lead to the need for a fresh approach towards measuring its effectiveness as well. Without the relevant content measurement methodologies businesses can make flawed strategic decisions, wasting resources and sending the wrong brand signals. As Warren Buffet famously quoted, "It takes 20 years to build a reputation and five minutes to ruin it. If you think about that, you'll do things differently."

Measuring Content Beyond Traditional Metrics

Measuring content's impact and brand value is crucial in today's economy, where roughly 80% of the S&P's value comes from intangible assets. As accounting standards potentially evolve to incorporate branded assets on corporate balance sheets, the imperative question arises: How can the value and impact of content be effectively measured?

As with any other modern and effective measurement framework, when measuring content effectiveness, it's important to align metrics with your specific strategic objectives. Strategic alignment ensures every measurement directly supports your organization's goals. Without this connection, teams may chase metrics that look impressive (or not), but don't influence business outcomes. Achieving effective content measurement requires a balanced approach to goals and results:

For awareness and reach objectives, track impressions, unique viewers, time spent with content, and social sharing metrics: These indicators show how widely your content is circulating and whether it's capturing audience attention.

Engagement metrics provide deeper insight into audience connection, which can be determined by monitoring comments, shares, likes, saved content, time spent, and completion rates (particularly crucial for video content). High engagement typically signals content resonance and can predict future brand affinity.

Audience sentiment analysis helps gauge how your content affects brand perception and the quality elements that lead to that. Track sentiment in comments, measure brand lift studies pre/post-content exposure, and monitor qualitative feedback to understand emotional responses. It's key to understand how consumers, channels, influencers, and all-important parts of the puzzle talk and feel about or engage with the brand.

Conversion-focused metrics connect content to business outcomes by measuring lead generation, newsletter signups, content downloads, and direct sales attribution. Track and analyze audience behavior, which ultimately matters the most as it is linked to what people do, and not just what people say they do.

Many organizations also use multi-touch attribution models that acknowledge content's role in the customer journey rather than expecting immediate conversion. This approach recognizes that content typically influences decisions rather than triggering immediate transactions and sees content as a long-term investment, not as a short-term advertising tactic.

Another way to understand content's impact is the in-depth analysis of visual content performance and the contribution of certain creative elements to audience engagement and brand perception. Understanding which creative ideas best communicate your brand's values and stories can transform your business. It drives higher conversions while optimizing costs and allows you to adapt content across different audiences, markets, and cultures.

Linking all of these important metrics and methodologies with financial performance finally concludes a good content measurement framework. Long-term brand health metrics like Net Promoter Score changes, customer acquisition cost reduction, and customer lifetime value provide a meaningful assessment of content effectiveness over time, and its contribution to brand economics.

Finally, it's worth mentioning that the right measurement framework should be easy to follow and adapt the same way business narratives adapt to cultural shifts or other external factors. An external factor of increasing significance is, of course, AI. AI has significantly redefined possibilities and boundaries when it comes to measuring content and brand value.

AI-Enabled Content Impact Measurement

The modern content ecosystem generates an immense volume and variety of data signals. This involves structured data (numerical metrics like views and clicks), semi-structured data (tagged social posts and categorized interactions), and unstructured data (comments, conversations, and visual engagement). Tracking, measuring, and managing the complexity of this data, but also analyzing it and triggering actions—off the back of the measurement process—and doing all this manually becomes a cumbersome task or series of tasks.

AI transforms the landscape by automating the entire measurement and analysis workflow. Machine learning algorithms can ingest massive datasets spanning multiple formats simultaneously, processing information that would otherwise overwhelm human teams. Natural language processing extracts sentiment and meaning from unstructured text at scale, while computer vision analyzes visual content engagement patterns.

AI systems can also identify correlations and patterns across seemingly unrelated data points, connecting content performance to business outcomes in ways human analysts might miss. These systems become increasingly accurate over time as they learn from each analysis cycle. The result is a shift from periodic, labor-intensive reporting to continuous, comprehensive measurement that provides deeper strategic insights while freeing human analysts to focus on interpreting results and developing business critical responses rather than processing raw data.

By leveraging AI, brands can move beyond superficial content metrics to truly understand and influence narrative effectiveness—measuring not just whether content is seen, but how deeply it resonates with their audiences. Traditional audience segmentation in marketing involves mostly basic demographics. With the introduction of AI, brands can identify more complex behavioral patterns across online touchpoints. AI detects audience micro-segments with specific needs and reveals connections between seemingly unrelated behaviors.

Further to this, machine learning empowers psychographic profiling and uncovers engagement drivers. Natural language processing extracts emotional responses from audience interactions, while sentiment analysis tracks evolving attitudes or cultural shifts. Through deep analysis of content preferences, AI can understand values rooted in audience decisions, and lead to content that talks to audiences' psychology beyond surface-level connection.

Finally, the full potential of AI is realized in the nuanced and precise execution of content personalization strategies. AI enables dynamically assembled content based on individual signals and needs, delivering content in the right place at the right time.

When AI is properly implemented, all these capabilities can create a virtuous cycle: Deeper understanding creates more relevant content, generating richer engagement data, which further refines understanding. This cycle helps build authentic relationships while delivering measurable business results.

Business Growth Through AI-Powered Content Analytics

Content analytics and "intelligent" content assisted by AI have the potential to directly accelerate business growth by transforming content from a cost center or a reporting function into a revenue-generating engine. AI attribution models can associate content directly with revenue, identifying exactly which content assets and channels drive sales. Companies that understand and implement these models experience higher ROI, reallocating budgets to high-converting content types directly impacting top-line growth.

Also, by analyzing engagement patterns across content portfolios, AI identifies underserved audience segments with high conversion or growth potential. Organizations leveraging these insights expand their addressable markets significantly, fueling growth beyond the existing customer base.

Another aspect and significant growth driver is the alignment of brand narratives with brand loyalty and brand advocacy. By analyzing and understanding which story elements or creative ideas best communicate their brand values, businesses can drive higher emotional engagement and sharing behaviors. Brand advocates represent a strong revenue pillar but also bring additional customers through word of mouth, creating a powerful growth multiplier effect.

In addition to being a revenue booster, AI-powered content analytics contribute to operational efficiencies and significant cost savings. Being able to predict and test the performance of content and creative ideas allows brands to allocate resources intelligently and efficiently. Also, automating workflows and processes with AI allows for leaner and more flexible organizational structures. This enables effective budget management and drives further profitability gains. These gains enhance overall business profitability or fuel brand expansion through reinvested savings.

The business impact of AI-powered content is substantial and measurable: Organizations effectively implementing content analytics typically achieve two to three times greater revenue growth from their content investments compared to traditional approaches. Equally important, they establish content as a predictable, scalable growth driver with direct impact on revenue, market share, and profitability.

Building an AI-Powered Content Analytics Framework

But what actions are required of businesses to realize this growth trajectory and build these innovative AI-enabled content analytics capabilities?

Defining AI-Measurable Business and Content Objectives

It all starts with clear and realistic goal-setting. Convert high-level business goals (revenue growth, market share, and customer loyalty) into specific content performance indicators and define quantifiable metrics for each objective with defined thresholds for success. As part of this step, it's important to align content performance with expected business outcomes. This action will further drive objective hierarchies and measurement taxonomies for consistent content classification—types, formats, and journey stages.

Defining the Strategy to Meet These Objectives

Strategy definition is a critical component for any framework in order to have a sense of direction. Strategy articulates how you'll achieve those goals—the approach, resources, and methods you'll employ. It dictates the roadmap and specific actions that will help you reach your destination.

In an effective AI-powered content analytics framework, the AI and content strategy must be fully aligned with the overall business strategy. Further to the strategy definition, audience targeting priorities and content distribution channels have to be clearly laid out. Other parameters to be included in this step include differentiation principles against competitors, resource allocation strategy across content types and platforms, as well as guidelines for balancing short-term performance with long-term brand building. It's key to establish a comprehensive decision-making loop where both goals and strategies can be refined based on AI-generated insights.

Identifying and Implementing Appropriate AI Technologies

This step will incorporate the necessary AI components as dictated by the defined strategy. While many organizations begin by acquiring tools and subsequently seeking their purpose and successful application, the establishment of a strategic framework prior to technology identification is a key differentiator.

Creating an AI-Ready Data Infrastructure

As we know, there is no AI without data. To enable advanced AI applications, businesses need a solid data foundation. When it comes to content analytics, this includes both the actual data management strategy and tools, but also clarity on the data sets to capture, store, and work with.

Having a robust and scalable data warehouse and data lake for centralizing both structured and unstructured data is of paramount importance. Technologies such as data clean rooms that enable secure data sharing and cross-party collaboration without physically moving data are becoming very popular as well. Cross-platform tracking with consistent identification parameters, first-party data collection processes compliant with privacy regulations, automated data validation and standardized formatting protocols, and the creation of unified data models connecting content, audience, and business metrics are essential steps from a data readiness perspective.

Developing AI-Assisted Use Cases

The value of data and AI is contingent upon the realization of specific business use cases. This step entails the application of data and AI to practical business solutions and use cases. Such use cases can be the configuration of personalized dashboards and self-service capabilities for higher business adoption and engagement, automated insights generation, development of predictive or prescriptive models for content performance forecasting, development of recommendation and decision-making engines, and implementation of attribution analysis connecting content touchpoints to the top line.

Implementing AI-Powered Feedback Loop

To successfully complete the process of setting up content analytics you need a continuous feedback loop that will enable learnings and the refinement of the original approach. Implementing closed-loop analytics, connecting performance metrics to content creation, and building AI performance prediction models will provide the required validation and valuable insights. It's key for organizations to ensure that what they built drives ongoing improvement.

This framework creates a sophisticated and comprehensive AI-powered content analytics system that measures and links content performance and effectiveness with business performance and growth. The success depends on balancing technical implementation with alignment to strategic business objectives.

Challenges to Consider in AI-Driven Content Analytics

AI systems can perpetuate biases in their training data, potentially favoring certain content styles while undervaluing others. Organizations must implement bias detection through regular audits, balanced reference datasets, and continuous monitoring across audience segments. Leading frameworks now incorporate fairness constraints directly into algorithms and conduct regular bias impact assessments to prevent systematic disadvantages to certain voices.

The granular data needed for AI analytics sometimes clashes with privacy regulations like GDPR and CCPA. Organizations should implement privacy-by-design principles including data minimization and robust anonymization. Transparency mechanisms with clear disclosure about measurement methods and opt-out processes are essential. Some companies now employ federated learning where models train across distributed data without centralizing sensitive information.

Another important aspect is over-reliance on AI optimization risks creating formulaic "Frankenstein" content that seems to maximize metrics at the expense of creative storytelling. Successful organizations establish boundaries between creative development and optimization, using AI as feedback rather than direction. Others implement "optimization-free zones" for experimentation without immediate performance pressure. The most effective approach positions AI as a creative collaborator rather than replacement of human creativity.

Additionally, complex "black box" models can deliver accurate predictions, but also offer limited visibility into their reasoning or decision-making processes. This creates challenges for stakeholder trust and effective collaboration. Organizations must implement interpretable AI approaches, including featuring the importance of analysis and choosing transparent systems over complex ones.

Finally, forward-thinking companies acknowledge the need to upskill their human capital and invest in literacy programs to help teams understand AI capabilities and limitations. This helps tremendously in terms of closing any potential knowledge gap but also ensuring there is an educated human in the loop.

About the Author

Nikolaos Lampropoulos is an advisor and entrepreneur specializing in AI and data analytics applications primarily within the media and creative industries. With a unique blend of business acumen and technical expertise, he helps forward-thinking organizations navigate digital transformation and unlock new growth opportunities.

His work focuses on building bespoke business strategies that leverage advanced analytics capabilities, AI-powered solutions, and innovative approaches that generate new business revenue streams for his clients. As a trusted consultant to executive teams, he helps organizations explore new business models that leverage the transformative power of data and AI.

With a growth mindset and cross-disciplinary curiosity, Nikolaos is an industry thought leader and speaker, bringing intellectual rigor, pragmatism and authentic enthusiasm to future-shaping and industry-defining conversations.

Email: nik@shapesandnumbers.com

Website: www.shapesandnumbers.com

CHAPTER 16

AI LITERACY AND THE NEXT GENERATION OF DIGITAL THINKERS

By Anastassia Lauterbach, PhD
CEO AI Edutainment, Professor for AI
Basel, Switzerland

*We choose to go to the Moon in this decade and do the other things,
not because they are easy, but because they are hard; because that
goal will serve to organize and measure the best of our energies and
skills, because that challenge is one that we are willing to accept, one
we are unwilling to postpone, and one which we intend to win.*
—John F. Kennedy, Rice University Speech, 1962

Why AI Literacy Matters Now More Than Ever

In May 2025, I attended a conference with public sector representatives. One optimistic AI keynote from an Oxford professor showcased self-driving cars on the streets of Los Angeles and San Francisco.

The audience cheered and gasped while the professor discussed the abolishment of human labor in the new wave of techno-optimism. Ministers from different countries applauded the autonomous driving technologies, believing they were reliable and mature and a minute away from being implemented outside of California. Nothing could be further from the reality. Bank of America estimates that Alphabet, the parent company of Waymo, spends approximately USD 1.5 billion per year on implementation of self-driving technologies, which translates to roughly USD 375 million per quarter.[1] Automated vehicles may not ensure a break-even point in other companies.

General Motors (GM), which invested heavily in its Cruise autonomous vehicle subsidiary, has spent between USD 10 billion and USD 16 billion since acquiring a controlling stake in 2016. The lower end of this range—over USD 10 billion—reflects direct operating losses. At the same time, the higher figure likely includes broader capital expenditures and acquisition costs.[2] The high cost per vehicle, combined with regulatory setbacks and public trust issues, led GM to abandon Cruise's robotaxi ambitions and refocus on integrating autonomous technology into personal vehicles.

I have a lot of empathy for companies investing millions and billions into innovation. I believe we must discuss the approaches, for example, asking, is it wise to put big beds on sensors that can get broken or dirty and impair the data flow? Are synthetics the only options to increase the volume of data for modeling purposes? Were Musk and Karpathy right to deliver Tesla cars equipped with streaming video capabilities? The point is not in discussing autonomous vehicles and their expensive road to profitability. It is about a healthy questioning of the professor's remarks. I looked at his profile after the conference. He teaches religion, not AI, computer engineering, or statistics. Sometimes, a well-intentioned attempt to inspire an audience can bring more confusion than anticipated. This is highly unfortunate, as in the next few decades, the humanities will need more data technologies than ever before.

The integration of artificial intelligence and data literacy into modern life is essential due to AI's transformative impact on scientific advancements, economies, and societal challenges, such as demographic decline. In this context, we require a balanced and critical view on how to forecast the benefits and downsides of AI technologies, not to downplay them but to prevent escalated valuations and unnecessary anger from investor communities, shareholders of incumbent companies, and philanthropists donating their money for AI research and development.

As a global society, we require AI to progress. AI models like AlphaGeometry solve complex mathematical problems, advancing fundamental research.[3] Robots and AI fill gaps in manufacturing, healthcare, and elder care, particularly where middle-aged workers are in short supply.[4]

Over the past 15 years, digital assets—including data, IT infrastructure, and intellectual property (IP)—have become central to company valuations. The rise of technology-driven business models and the digital transformation of traditional industries have shifted value creation from tangible to intangible assets at an unprecedented scale. By 2023, intangible assets—including digital assets and IP—accounted for approximately 90% of the enterprise value of the top 15 US companies.[5]   AI is a catalyst for the rapid proliferation of digital assets, enabling their automated creation, streamlining management, and enhancing discoverability. Security is a two-edged sword, as cybercriminals utilize AI technologies to successfully steal data and intellectual property and disrupt the supply and value chains of countless businesses. On the other hand, applying AI in corporate defenses bolsters security. Its integration into digital asset management systems transforms these assets from mere storage solutions into strategic business drivers, unlocking new efficiencies and opportunities for innovation across industries.

A lack of AI literacy can have severe consequences when it comes to determining how to regulate emerging technologies. The merging of the state and private sectors, the struggles of democracy, and data protection will remain weak excuses for why the European technology ecosystem cannot keep pace. As overregulation has become the main factor influencing competitiveness, European-based businesses will continue to struggle with escalating costs.

Place yourself in the shoes of an AI mid-sized business in Europe or the US. In best case, this business is already spending 25% of the revenue line on cloud, and a further 15% on cleansing data, before any training of an AI model is initiated.[6] What might seem feasible for Big Tech players in terms of operating expenses and investments isn't digestible for startups and mid-sized innovators. As of 2023, Alphabet holds USD 118 billion in cash reserves.[7] Meta Platforms reported USD 70.23 billion in cash, cash equivalents, and marketable securities as of March 31, 2025.[8] According to a survey cited by proALPHA, 34% of European manufacturing companies are planning to relocate production due to the EU AI Act, compared to an industry average of 26%.[9] The act is perceived as a substantial barrier to innovation and international competitiveness, particularly in Germany and other industrial hubs. As a result, Europe won't experience the growth badly needed to attract talent from abroad and maintain the existing social order.

I chose Kennedy's quote to open the article on AI literacy as it captures the spirit of tackling daunting, multifaceted challenges for the sake of progress. Introducing AI literacy across society is a modern parallel to Kennedy's moonshot—a complex, demanding, but essential mission.

What Is AI Literacy?

The "AI literacy" concept encompasses a multi-dimensional set of competencies that go beyond mere understanding of technologies and engineering. It requires critical thinking, ethical reasoning, and practical application—paralleling the expanded role of literacy in a digital and information-driven society.[10] The AI Literacy Framework (AILit), developed by the European Commission and the Organisation for Economic Co-operation and Development (OECD), defines AI literacy as a blend of knowledge, skills, and attitudes. These frameworks provide practical domains and competencies for engaging with, creating with, managing, and designing AI systems.[11] A well-crafted definition of artificial intelligence is crucial for facilitating understanding of these technologies, both for technical experts and the general public.

AI today is broadly defined as data technology that mimics human capabilities and decision-making properties in order to forecast, simulate, augment, and advise. This definition helps people distinguish AI from traditional software or automation, highlighting its unique ability to mimic, augment, or extend human cognitive functions.

Some AI definitions include the emulation of human learning in machines. However, human learning could not be further from how these technologies work due to the properties and evolution of human brains. The main issue is the ability of humans to learn from counterfactuals.

Children use pretend play to explore counterfactuals, which correlates with causal reasoning skills.[16] This suggests an evolutionary advantage: Extended immaturity in children before they reach adulthood allows for experimentation with alternative outcomes, thereby refining causal models.

Humans identify root causes by contrasting reality with imagined alternatives (e.g., "If I hadn't missed the train, I'd be on time"[12]). Counterfactuals help predict outcomes of future actions (e.g., avoiding risky choices after imagining negative consequences[13]). Finally, mental simulations allow testing hypothetical actions without real-world risks.[14]

Yann LeCun, Meta's Chief AI Scientist, has stated: "We are really far from human-level intelligence... The largest LLMs have about the same number of parameters as the number of synapses in a cat's brain... So maybe we are at the size of a cat. But why aren't those systems as smart as a cat?... We are nowhere near that. We are still missing something big... A cat can remember, can understand the physical world, can plan complex actions, can do some level of reasoning—actually much better than the biggest LLMs."[15]

Policymakers address (historical) biases in datasets, lack of fairness criteria in modeling parameters, and the mistakes/hallucinations an AI can make. They neglect the fact that these flaws can't be eliminated, even if researchers and practitioners introduce sophisticated error-resilience frameworks in transformers, expand datasets with synthetics to improve model generalization,[16] or play with the so-called temperature criteria that make LLMs sound more human.[17]

Transformers and deep learning models are fundamentally prone to hallucinations due to architectural constraints that cannot be fully resolved within their current paradigms. Regulators would do well to find ways to incentivize work on the algorithmization of causality, hybrid architectures, and investing in science around explainable AI. The US National Science Foundation (NSF) allocates approximately 10% to 15% of its annual AI research funding to "trustworthy AI" research, which includes interpretability, robustness, privacy preservation, and fairness. Within this, interpretability and explainability (XAI) receive about 2% of total AI funds, robustness about 6%, and privacy preservation has risen to as much as 5% in recent years.[18] The remaining 85% to 90% of NSF AI funding is directed toward general AI capabilities and applications rather than causality research, safety, or explainability. In these circumstances, the road to alternative architectures in AI may be very long.

Engaging with differences in learning between mammals (including humans) and machines would serve us well when it comes to defining AI literacy. Current frameworks don't address it, ensuring that educators and regulators work on symptoms rather than addressing the root cause.

AI Literacy in Educational Systems

Only a few governments have elevated AI literacy into the stardom of being a "must have" in their education systems. Since the 2023 to 2024 academic year, Hong Kong has mandated 10 to 14 hours of AI education for junior secondary students (forms one to three). The curriculum includes lessons on basic AI concepts, computer vision, speech and language understanding, robotic reasoning, AI ethics, and social impacts.[19]

The UAE is the first country to make AI a mandatory subject across all school grades, from kindergarten through grade 12, in all government schools starting from the 2025 to 2026 academic year.[20] China will follow this path starting September 1, 2025.[21] The victory of AlphaGo over Lee Sedol in March 2016 profoundly reshaped China's perception of artificial intelligence and triggered changes in its educational system. This event, watched by over 280 million Chinese viewers,[22] served as a technological "Sputnik moment," igniting national urgency around AI development, and it gave AI scientists and practitioners the status of rockstars.[23]

Interestingly, initiatives like the European Commission and OECD's AI Literacy Framework aim to make AI literacy a core competency in education systems worldwide,[24] but they have failed to influence practical implementation of AI literacy in Belgium and France, countries where the headquarters of the European Commission and OECD are located. While governments consider including AI literacy into national school programs, two camps of "yey-" and "nay-" sayers emerge, either celebrating the use of AI and generative AI in education, or campaigning against these technologies.

The critics of AI emphasize the irreplaceable role of teachers, the risks of algorithmic bias, and the need for policies that prioritize human instruction over automated solutions.[25] Overreliance on LLMs discourages independent analysis and reduces opportunities for students to develop original thought.[26] A Microsoft and Carnegie Mellon study claims a self-reported decline in confidence and critical thinking abilities among knowledge workers due to the use of generative AI (GenAI).[27] Delegating tasks to AI leads to "cognitive debt," where deferred mental effort results in long-term skill atrophy.[28]

Based on current research, there is no universally agreed-upon timeframe to definitively determine long-term cognitive decline due to AI, as studies to date have shown correlative, not causal, relationships over relatively short periods.[29] One of the few studies on ChatGPT in education provided plausible results, showing that generative AI enhanced task-specific outcomes and engagement but had a limited impact on more profound learning, such as critical thinking and analysis. While students felt more motivated when using ChatGPT, this did not translate into better long-term knowledge retention or deeper understanding.[30]

Another critical study reviewed the use of large language models in programming. LLMs improved task performance during assignments but significantly reduced students' ability to solve similar problems independently in controlled settings. The findings facilitate rethinking of when we introduce generative AI into coding classes. While exposing beginners will negatively impact their skill development, advanced programmers can benefit from AI as a collaborative tool.[31]

When it comes to math, tailored generative AIs like CEMAL can help train students to achieve good performance through a customized feedback loop and adjustments tailored to individual strengths and weaknesses.[32] Though I empathize with the frustration of the academic community, I wish for an elaborate approach. Young adults won't understand the "how" behind human and machine collaboration unless they spend time learning the underlying foundation of AI technologies, their history, and their limitations.

A Novel Approach to Enhance AI Literacy

Components of AI Literacy include four major themes.

1. Start early, and edutain about AI while using storytelling based on real science.

"Edutainment" is a blended approach that combines educational content with entertaining elements, aiming to make learning both effective and enjoyable. The term itself is a portmanteau of "education" and "entertainment." It has been used to describe a wide range of media, experiences, and methodologies designed to teach while keeping learners engaged.

My own flagship project is the *Romy & Roby* book series, starting with *Romy, Roby and the Secrets of Sleep* (www.romyandroby. ai). I utilize storytelling to introduce children to AI concepts in a natural and relatable way. The stories follow a family and their speaking robot, Roby, who makes mistakes typical of today's AI (like confusing landmarks) and learns from them—mirroring real-world AI learning processes. The books are designed not just for children but also to facilitate discussions among teenagers and adults, bridging generational gaps in technology understanding and increasing parental involvement.

2. *Review the development of AI literacy in relation to how we facilitate human skills and cognition in general.*

While discussing AI education, I can't forget two areas of knowledge that greatly contribute to enhancing human critical thinking and the ability to communicate. I encourage schools and universities in any country to compulsorily teach students of all age groups two subjects—(creative) writing and applied mathematics.

The first stimulates critical thinking and encourages students to separate original ideas from cheap prompting. William Faulkner reportedly said something similar: "I don't know what I think until I read what I said," or "I never know what I think about something until I read what I've written." Stephen King also expressed a similar sentiment in 2005: "I write to find out what I think." I could go on to find historical evidence on the value of writing, but the point is clear. Any professional can benefit from the clarity of thought as it evolves through serious training in writing skills. I can't applaud Jennifer A. Doudna enough, not just for winning the 2020 Nobel Prize in Chemistry but for popularizing her research while writing *A Crack in Creation*[33] or David Deutsch's work in quantum physics that he put on display for everyone who cares in *The Fabric of Reality*[34] and *The Beginning of Infinity*.[35]

Teaching applied mathematics to everyone isn't just about popularizing numerical skills. It is about reducing fears that something as abstract as equations might be fun and valuable. Reading statistics is great in every profession, enabling a better understanding of the fundamental technologies powering AI. When discussing applied mathematics, I don't invite the world to reinvent the wheel. The International Baccalaureate (IB) Diploma Programme has been offering a course, "Mathematics: Applications and Interpretations" (MAI) for quite some time now, greatly benefiting those students who seek future degrees in medicine, psychology, or the arts.

3. *Support community leadership to enable AI literacy.*

Community leadership can play a crucial role in advancing AI literacy by mobilizing resources, promoting inclusive education, and fostering trust and engagement between AI companies and startups, schools, universities, colleges, and other organizations (e.g., NGOs) at the local level.

As an example, Lucerne has a local AI and cognitive research community, LAC2, which organizes breakfasts with AI researchers and developers, building bridges between the Lucerne University of Applied Sciences and Arts and the regional industry and working with the city administration to help startups connect and network. Adding schools into the equation will increase the impact of the initiative.

4. *Shape a dialogue on how AI and robotics are portrayed in literature and film, as these portrayals influence public perception of the technologies.*

The fifth component is about developing a vision of how humans and AIs might interact once AI is omnipresent and more sophisticated. A while ago, I started talking about portrayals of AI and robotics in science fiction on my podcast, *AI Snacks with Romy&Roby.*

Science fiction constantly probes philosophical questions about what it means to be human in an AI-augmented world. It can significantly help us anticipate and navigate the psychological and cultural shifts AI might bring. Remember *Klara and the Sun* and the dilemma of what might constitute the ultimate use case for a robot—replacing a sick child?[36] Remember Data from *Star Trek* and his attempts to humanize himself?[37] Those are edutainment attempts, not classical entertainment. In this context, I will dedicate part of my time in 2025 and 2026 to recording several podcast episodes on AI in books and movies and feature books such as the *Three-Body Problem* by Cixin Liu,[38] *Machines Like Me* by Ian McEwan,[39] books by Stanisław Lem, Arkady, Boris Strugatsky, and Isaac Asimov, films like *Her*,[40] *Ex Machina*,[41] and *Chappie*,[42] and series like *Westworld*.[43]

AI and robotics are here to stay; our economies rely on innovative data technologies. We cannot outsource complete control over our future with AI to a few technology giants from the US and China. No one has the resources to overspend on piloting machine learning tools because we can't differentiate between real expertise and AI wannabes. The next generation of digital thinkers should encompass people from diverse age groups and backgrounds to ensure we benefit from AI technologies.

AI literacy isn't a one-way street. It is a comprehensive ecosystem of topics and ideas open to new businesses to form and traditional companies to engage with and thrive in. The endgame for democratizing knowledge of AI and robotics hasn't yet begun. It won't be played optimally if the broader public doesn't raise its voice on what it expects from its future with AI and robots. Edutainment is a powerful tool for engaging both youth and adults in AI. Community leadership is crucial for the implementation of AI literacy. Ultimately, even in the realm of technology, the human factor is everything.

References

1. Grayson Brulte, "$50 Billion to Create Metaverse, $30 Billion to Create Waymo," *The Road to Autonomy*, https://

www.roadtoautonomy.com/metaverse-waymo-spending/ (accessed June 30, 2025).

2. "General Motors to Retreat From Robotaxis and Stop Funding Its Cruise Autonomous Vehicle Unit," *The Economic Times*, December 11, 2024, https://economictimes.indiatimes.com/news/international/business/general-motors-to-retreat-from-robotaxis-and-stop-funding-its-cruise-autonomous-vehicle-unit/articleshow/116191669.cms(accessed June 30, 2025); Stephen Council, "Cruise, SF's Embattled Self-Driving Car Company, Is Finally Folding After $10B in Losses," *SF Gate*, December 10, 2024, https://www.sfgate.com/tech/article/sf-robotaxi-company-cruise-folds-gm-cuts-funding-19972259.php (accessed June 30, 2025).

3. Google Keyword Team, "9 Ways AI Is Advancing Science," *The Keyword* (blog), 2024, https://blog.google/technology/ai/google-ai-big-scientific-breakthroughs-2024/ (accessed June 26, 2025).

4. Daron Acemoglu and Pascual Restrepo, "Demographics and Automation," 2021, https://pascual.scripts.mit.edu/research/demographics/demographics_automation_restud.pdf (accessed June 26, 2025).

5. Sarath, "Intellectual Property Valuation: Case Study," *Eqvista*, 2024, https://eqvista.com/intellectual-property-valuation-case-study/ (accessed June 26, 2025).

6. Martin Casado and Matt Bornstein, "The New Business of AI (and How It's Different from Traditional Software)," *Andreessen Horowitz*, 2020, https://a16z.com/the-new-business-of-ai-and-how-its-different-from-traditional-software/ (accessed June 26, 2025).

7. OfficeChai Team, "The Cash Reserves of These Top Tech Giants [2023]," *OfficeChai*, 2023, https://officechai.com/stories/the-cash-reserves-of-the-top-tech-giants-2023/ (accessed June 26, 2025).

8. "Meta Platforms Cash on Hand 2010–2025," *Macrotrends*, 2025, https://www.macrotrends.net/stocks/charts/META/meta-platforms/cash-on-hand (accessed June 26, 2025).

9. M. Finkler, "Supply Chain Act and EU AI Act Slow Down the Manufacturing Industry," *proALPHA* (blog), 2024, https://www.proalpha.com/en/blog/ai-act-eu-regulation-smes-report (accessed June 26, 2025).

10. Gilliam Writers Group, "Beyond Reading and Writing: The Expanding Role of Literacy in a Digital World," *Gilliam Writers Group*, 2025, https://www.gilliamwritersgroup.com/blog/beyond-reading-and-writing-the-expanding-role-of-literacy-in-a-digital-world-43zk4 (accessed June 26, 2025).

11. K. Mills et al., "AI Literacy: A Framework to Understand, Evaluate, and Use Emerging Technology," *Digital Promise*, 2024, https://doi.org/10.51388/20.500.12265/218; https://digitalpromise.org/2024/06/18/ai-literacy-a-framework-to-understand-evaluate-and-use-emerging-technology/ (accessed June 26, 2025).

12. N. Van Hoeck et al., "Counterfactual Thinking: An fMRI Study on Changing the Past for a Better Future," *Social Cognitive and Affective Neuroscience* 8, no. 5 (2012), https://doi.org/10.1093/scan/nss031; https://pmc.ncbi.nlm.nih.gov/articles/PMC3682438/ (accessed June 26, 2025).

13. N. Van Hoeck et al., "Counterfactual Thinking: An fMRI Study on Changing the Past for a Better Future," *Social Cognitive and Affective Neuroscience* 8, no. 5 (2012), https://doi.org/10.1093/scan/nss031; https://pmc.ncbi.nlm.nih.gov/articles/PMC3682438/ (accessed June 26, 2025).

14. D. Matovski, "Causal AI: The Revolution Uncovering the 'Why' of Decision-Making," *Global Policy* (blog), April 23, 2024, https://www.globalpolicyjournal.com/

blog/23/04/2024/causal-ai-revolution-uncovering-why-decision-making(accessed June 26, 2025).

15. S. Cao, "Meta's A.I. Chief Yann LeCun Explains Why a House Cat Is Smarter Than the Best A.I.," *Observer*, February 2024, https://observer.com/2024/02/metas-a-i-chief-yann-lecun-explains-why-a-house-cat-is-smarter-than-the-best-a-i/(accessed June 26, 2025).

16. Exxact, "7 Common Machine Learning and Deep Learning Mistakes and Limitations to Avoid," *Exxact Blog*, 2023, https://www.exxactcorp.com/blog/Deep-Learning/7-Common-Machine-Learning-and-Deep-Learning-Mistakes-and-Limitations-to-Avoid (accessed June 26, 2025).

17. ChatBotKit, "How to Prevent AI Model Hallucinations," *ChatBotKit*, https://chatbotkit.com/tutorials/how-to-prevent-ai-model-hallucinations (accessed June 26, 2025).

18. L. Alexander and D. Kaushik, "Trust Issues: An Analysis of NSF's Funding for Trustworthy AI," *Federation of American Scientists*, 2023, https://fas.org/publication/trust-issues/ (accessed June 26, 2025).

19. Bangbangtax, "Mandatory AI Curriculum in Schools: China, the US, and the Global Race," *Banthetax*, 2025, https://banthetax.com/mandatory-ai-in-schools/ (accessed June 26, 2025).

20. S. Shery, "UAE Becomes First Country to Make AI a Mandatory Subject for All School Grades," *TEC Spectrum*, 2025, https://tecspectrum.com/happenings/uae-ai-education-mandatory-schools/ (accessed June 26, 2025).

21. ·Asia Education Review Team, China to Introduce Mandatory AI Education in Schools by 2025, *Asia Education Review*, 2025, https://www.asiaeducationreview.com/technology/news/china-to-introduce-mandatory-ai-

education-in-schools-by-2025-nwid-3573.html (accessed June 26, 2025).

22. Y. Dong, "AlphaGo and the Clash of Civilizations," *Foreign Policy*, March 18, 2016, https://foreignpolicy.com/2016/03/18/china-go-chess-west-east-technology-artificial-intelligence-google/ (accessed June 26, 2025); Kai-Fu Lee, "China's Sputnik Moment and the Sino American Battle for AI Supremacy," *Asia Society Magazine*, 2018, https://asiasociety.org/magazine/article/chinas-sputnik-moment-and-sino-american-battle-ai-supremacy (accessed June 26, 2025).

23. A. Lauterbach, "Trojanische Verhältnisse?," in T. Loitsch, ed., *China im Blickpunkt des 21. Jahrhunderts: Impulsgeber für Wirtschaft, Wissenschaft und Gesellschaft* (Springer, 2019), 1–17.

24. "Why AI Literacy Is Now a Core Competency in Education," *World Economic Forum*, 2025, https://www.weforum.org/stories/2025/05/why-ai-literacy-is-now-a-core-competency-in-education/ (accessed June 26, 2025); Yolanda Gil and Raymond Perrault, "Artificial Intelligence Index Report 2025," 2025, https://hai-production.s3.amazonaws.com/files/hai_ai_index_report_2025.pdf (accessed June 26, 2025).

25. J. Senechal et al., "Balancing the Benefits and Risks of AI Large Language Models in K12 Public Schools," *Virginia Commonwealth University—VCU Scholars Compass*, 2023, https://scholarscompass.vcu.edu/cgi/viewcontent.cgi?article=1133&context=merc_pubs (accessed June 26, 2025).

26. E. Harvey, A. Koenecke, and R. Kizilcec, "'Don't Forget the Teachers': Towards an Educator-Centered Understanding of Harms from Large Language Models in Education," *arXiv*, February 2025, https://arxiv.org/html/2502.14592v1(accessed June 26, 2025).

27. H. Lee et al., "The Impact of Generative AI on Critical Thinking: Self-Reported Reductions in Cognitive Effort and Confidence Effects from a Survey of Knowledge Workers," *CHI 2025*, https://doi.org/10.1145/3706598.3713778; https://www.microsoft.com/en-us/research/wp-content/uploads/2025/01/lee_2025_ai_critical_thinking_survey.pdf(accessed June 26, 2025).

28. R. Loga, "AI's Cognitive Implications: The Decline of Our Thinking Skills?," *IE University* (blog), 2025, https://www.ie.edu/center-for-health-and-well-being/blog/ais-cognitive-implications-the-decline-of-our-thinking-skills/(accessed June 26, 2025).

29. M. Meneses, "Thinking in the Age of Machines: Global IQ Decline and the Rise of AI-Assisted Thinking," *The Quantum Record*, 2025, https://thequantumrecord.com/philosophy-of-technology/global-iq-decline-rise-of-ai-assisted-thinking/(accessed June 26, 2025).

30. Y. M. E. Heung and T. K. F. Chiu, "How ChatGPT Impacts Student Engagement from a Systematic Review and Meta-Analysis Study," *Intelligence* 8 (2025), https://doi.org/10.1016/j.caeai.2025.100361; https://www.sciencedirect.com/science/article/pii/S2666920X25000013 (accessed June 26, 2025).

31. G. Jošt, V. Taneski, and S. Karakatič, "The Impact of Large Language Models on Programming Education and Student Learning Outcomes," *Applied Sciences*, 2024, https://doi.org/10.3390/app14104115; https://www.researchgate.net/publication/380583693_The_Impact_of_Large_Language_Models_on_Programming_Education_and_Student_Learning_Outcomes (accessed June 26, 2025).

32. Z. Liang et al., "Let GPT Be a Math Tutor: Teaching Math Word Problem Solvers with Customized Exercise Generation," *arXiv*, 2023, https://doi.

org/10.48550/arXiv.2305.14386; https://arxiv.org/abs/2305.14386 (accessed June 26, 2025).

33. Jennifer A. Doudna and Samuel H. Sternberg, *A Crack in Creation: Gene Editing and the Unthinkable Power to Control Evolution* (Mariner Books, 2017).

34. David Deutsch, *The Fabric of Reality: The Science of Parallel Universes and Its Implications* (Viking Adult, 1997).

35. David Deutsch, *The Beginning of Infinity: Explanations That Transform the World* (Penguin Books, 2012).

36. Kazuo Ishiguro, *Klara and the Sun* (Faber and Faber, 2021).

37. Gene Roddenberry, creator, *Star Trek* [TV series and films], 1966–present, https://www.startrek.com (accessed June 26, 2025).

38. Liu Cixin, *The Remembrance of Earth's Past* trilogy (Chongqing Publishing Group, 2008–2010).

39. Ian McEwan, *Machines Like Me* (Jonathan Cape, 2019).

40. Spike Jonze, director and writer, *Her* [film] (Warner Bros. Pictures, 2013).

41. Alex Garland, director and writer, *Ex Machina* [film] (A24, Universal Pictures International, 2014).

42. Neill Blomkamp, director and writer, *Chappie* [film] (Sony Pictures, 2015).

43. Jonathan Nolan and Lisa Joy, creators, *Westworld* [TV series] (HBO Entertainment, 2016–2022).

About the Author

Dr. Anastassia Lauterbach-Lang is the CEO and founder of AI Edutainment GmbH (www.aiedutainment.ai). This Switzerland-based company brings knowledge of AI, robotics, and quantum computing to one million families globally. Anastassia is the creator of *the Romy and Roby* AI edutainment series (www.romyandroby. ai), a professor of cybersecurity, AI, and business ethics in Innsbruck and Berlin, and a mentor to C-level executives in companies such as Pfizer, IBM, Alphabet, and SAP. Her podcast, *AI Snacks with Romy & Roby*, translates AI and robotic technologies from difficult to easy to understand.

Anastassia has extensive corporate governance experience and was a member of supervisory and advisory boards in companies such as Star Alliance, Intel, Dun & Bradstreet, and easyJet.PLC. She created the German and Austrian chapters of the Women Corporate Directors Foundation. For her actions in revealing fraud at Wirecard and her work in introducing cybersecurity and data technology-related governance frameworks on international boards, she was inducted into the global Hall of Fame for Business Excellence at the Business Excellence Institute in November 2024. Anastassia serves as an advisor to the Israeli Association on Artificial Intelligence and Ethics and the UN Global Mental Health Task Force. Anastassia is a former SVP and EVP with Qualcomm Inc., Deutsche Telekom AG, and T-Mobile International. She holds a master's degree in computational linguistics and psychology, and a PhD in linguistics. Anastassia is fluent in six languages and lives in Basel, Switzerland.

Email: anastassia@aiedutainment.ai

CHAPTER 17

FROM BLACK BOX TO GLASS BOX: BUILDING TRUST AND PROTECTING PERSONAL DATA IN THE AGE OF AI

By Bettina S. Lippisch
Intelligent Enterprise Transformation, AI Governance
Chicago, Illinois

It is up to humans to deploy AI responsibly and design a safe and ethical AI ecosystem that will ultimately help humans help themselves. Transparency can help demystify AI through chain of thought reasoning capabilities, information source attribution and/ or accountability for uncertainty. Explaining the "why" behind AI's behavior can help humans trust AI more.

—Veronika Rockova, Professor of Econometrics and Statistics, Booth School of Business of the University of Chicago

Hello There! Are You a Human?

When you pick up the phone to call your cable company or reach out to your health insurance via website chat, chances are, your first interaction is no longer with a human. The friendly voice of "Cynthia" or the helpful chat agent who introduced themselves as "Bobbie" might not be a human after all. Machine learning (ML) models-powered AI systems are rapidly evolving. This includes multi-layered neural networks that are based on the human brain and which can make complex, nonlinear inferences on vast amounts of unstructured data, like speech, text, and video. These systems can imitate human interactions in a way that is almost impossible to differentiate from a real person.

Advanced AI abilities, like computer vision (e.g., facial and pattern recognition), speech recognition (e.g., identity verification, sentiment detection) and natural language processing (e.g., chat bots), are now being further combined into large multimodal models (LMMs) and used in agentic AI, a fast-growing technology stack that combines AI, large language models (LLMs), automation, and machine learning into an autonomous, highly intelligent systems that can solve complex multi-step tasks without human interaction. And unlike human-to-human interactions, where the retention of information is limited to memory recollection or active documentation, machine-to-human interactions generate detailed and sustained sets of data, including personal information, often without consent or transparency about its future use.

So why does it matter if personal data is captured and used by machines? Imagine you are driving a car, and your insurance rate suddenly jumps by over 20%, even though you never had an accident. Of course you would shop around, but other insurance providers quote you the same high rates.

This is exactly what happened to a driver named Kenn in 2022. When he reached out to an insurance broker to ask why, he learned something unsettling: These rate hikes were based on a LexisNexis report, which contained hundreds of pages detailing his and his wife's use of their car, including start and end times, distances, sharp accelerations, hard braking events, and more, resulting in a "comprehensive driver risk score" report containing highly personal data. His insurance companies used it without his knowledge to determine the new insurance rates, and he was not alone.

Kenn's onboard AI system had been collecting his driving data and sold it to a third-party analytics firm, which in turn fed it into a machine learning model used by insurers. Kenn had neither spoken to a human at any point, nor did he ever consent to his personal data being shared with the analytics firm or the insurer. The terms of his car's tracking system did NOT disclose sharing his personal data with third parties, and he had no idea that the "conversations" that led to his higher premiums had been entirely between machines.

Privacy and Data Collection in the Age of AI

Kenn's story, brought to the attention of many privacy professionals and consumers alike through the research of *New York Times* journalist Kashmir Hill,[1] illustrates how privacy and data collection in the age of AI can be invisible, automated, and deeply consequential. Companies deploying AI systems may consider the commercial benefits but neglect to incorporate privacy by design/default, violating consumers' privacy rights as well as betraying their trust.

AI systems need data to work, and much of this data contains personal identifiable information, also referred to as PII. Yet, the models themselves are still very much "black boxes," even for those engineering them. This raises data privacy questions such as the following:

- *What happens inside the AI models?* There is an incredibly high velocity of data processing inside AI systems, and experts aren't able to fully explain and predict how these advanced models learn and evolve.

- *Who owns the data ingested, processed and generated by AI?* Consent and ownership of AI inputs and outputs, especially with generative AI, are hotly debated among creators, regulators, policy makers, and individuals. The speed of AI development has already outpaced legislation. Minimizing liability and risk today through privacy-enhanced, well-governed data should therefore be a priority for any organization BEFORE regulations and enforcements catch up.

- *How to govern PII, proprietary, or sensitive data within the AI model?* AI systems process enormous amounts of data, and the speed of processing makes it virtually impossible to extract or redact individual data sets after they have become part of an AI system. Privacy-enhancing technologies are evolving to address this, but at current state, the most feasible approach focuses on governing the inputs and outputs of a model along with strong privacy controls.

- *How is consent captured and honored?* Some legislation, like the EU AI Act,[2] already requires transparency when AI is present in automated processing and decision-making, but it is not (yet) a global requirement. As illustrated in Kenn's experience, consent mechanisms lag behind, deliberately or unintentionally, creating privacy harms and risks.

- *Who is accountable for its use?* While the EU AI Act is very specific about accountability, citing roles like developers, deployers, implementers, and users of AI, there is still a large question mark around accountability for AI systems and their data in other jurisdictions.

- *How to track data quality, lineage and origin when data is rapidly transforming in real time?* Traditional data quality and lineage approaches used in data governance frameworks do not fully cover AI systems. Once data enters into the model, it is instantly and consistently transformed. Until dynamic governance models catch up with AI systems, auditing output quality and testing against performance benchmarks during model development and use are the gold standard.

- *How do we govern decisions made by machines and track bias, misuse, or discrimination?* Bias, discrimination, and other harms during automated decision-making are some of the most common harms and risks in AI systems. Deploying privacy impact assessments (e.g., NIST's PIA[1]) can expose areas of high risk or blind spots to support thorough mediation.

Figure 1: AI Data & Privacy Considerations and Risks Related to Inputs and Outputs

But not just organizations deploying AI need to understand AI risks; end users also must comprehend how their personal data might be used for decisions made by machines. AI literacy is a critical aspect of AI governance, and its prominence in the EU AI Act under Article 4[3] is no accident.

---

[1]   6. National Institute of Standards and Technology (NIST), *AI Risk Management Framework (AI RMF)*, 2017, https://www.nist.gov/system/files/documents/2017/05/09/NIST-TIP-PIA-Consolidated.pdf (accessed September 13, 2025).

Some recent metrics related to AI systems illustrates why the floodgates of consumer pushback and related litigation could open quickly and soon for any company neglecting privacy considerations:

- *The AI-driven advertising market is projected to reach $2.2 trillion by 2030.*[4] Many, if not most of the first interactions consumers will have with companies and their products and services today will include some aspect of AI and rely on automated decision-making and profiling.

- *Global consumer trust in AI has fallen to 53%, with US trust at 35%.*[5] While the US still lags behind the global averages, considering the current political and economic climate and the rise in consumer awareness around data breaches and data misuse, organizations need to invest in trustworthy AI, or they will lose against competitors prioritizing transparent and ethical AI use.

- *Regulatory scrutiny on AI-powered targeting is intensifying globally. (EU AI Act (2024); Colorado AI Act (2024); FTC enforcement and guidance (2024–2025), etc.).* Similar to the GDPR, Europe's Data Protection Regulation, the EU AI Act is setting global standards for the responsible and transparent use of AI. Even though there is currently no comprehensive federal AI regulation in the US at the time of this writing, there are already state laws and other applicable regulations, including consumer protection, copyright and intellectual property (IP), and non-discrimination laws.

The Shifting Ground Beneath Privacy and Data Governance

All of the above have resulted in AI fundamentally altering the compliance landscape. Traditional privacy and data governance frameworks, which champion notice, consent, purpose limitations, transparency and policies, are no longer sufficient. While the same principles still mostly apply, the established methods become increasingly strained by the scale, speed, and opacity of AI systems.

The appetite of AI models for training data and the fragmented policy and regulatory ecosystem around consent, privacy, and consumer rights are creating the need to fundamentally rethink what it means to govern data privacy across AI systems. Where data was once collected for a specific purpose, and consent was given accordingly, most AI thrives on data maximization: the more data, the better the model and its outputs. Yet privacy laws often demand data minimization: Collect only what you need, for the purpose the individual has consented to. This creates a paradox. AI needs data to function, but privacy demands restraint.

Moreover, AI doesn't just automate decisions from data it ingests, it generates brand new data. It infers, predicts, and creates outputs that no longer track the data lineage back to the original data or consent. For example, AI models used in autonomous cars or trucks don't just learn about the driver and the car; they might collect data about the pedestrians, cyclists, and other drivers the car's sensors observed, and the personal and biometric data of strangers might end up in the model without them ever consenting to its use.

To add to the challenge in the AI universe, data is never truly static. It evolves, flows, and transforms, challenging traditional governance principles that assume data has a clear lifecycle and can be governed mostly at rest.

The Myth of Anonymity, Data "Fingerprints," and the Rise of Shadow Data

This constant velocity and processing power of AI systems has already transformed organizations, creating instant and highly customized user experiences, operational efficiencies, and rapid innovation. We already learned that there are often little to no indicators that an interaction is not with a human but with a machine, and the impact on data collection and privacy is significant.

Data generated by human-to-machine interactions is captured, stored, processed, augmented, and shared in detail and for potentially long periods of time. This rapid and often permanent storage of vast amounts of personal data across a multitude of systems and platforms is so granular that it can become so unique to a person, like a data "fingerprint," comparable to our DNA.

And while an individual interaction with an AI might not leave enough data to re-identify a person, the incredible speed and processing power of AI systems can take multiple data sets and find patterns and data points that can be matched to re-identifying a person, even if one dataset was successfully de-identified. Your phone's location history and your car's driving history may align so uniquely that someone could easily assume it is you and, therefore, re-identify you, even if one of the data sets was de-identified.

This ability to analyze large data sets with never-before-seen speed and accuracy can also be used to make assumptions about an individual based on circumstantial rather than explicit data. If someone has data sets containing info about your charitable donations, your income level, the car you drive, what city you live in, and which newspaper you read (all information available for many Americans from data brokerages or social media sites), AI systems able to processes enormous amount of socioeconomic, demographic, or behavioral data sets can potentially predict your political affiliation, immigration status, sexual orientation, or stance on political issues and append it to your record, even though you never directly provided this information willingly, or consented to its use.

This creation of Shadow Data undermines traditional privacy-protecting techniques and raises concerns about privacy harms and risks, if used for decision-making without the affected individual knowing about its existence or their ability to exercise their privacy or data subject rights.

Governing the Machine Mind

Both the velocity and volume of data available to and processed by AI requires a shift from static compliance and privacy checklists to dynamic, operational AI governance. This includes embedding privacy with security, data governance, risk management, and ethical considerations into the entire AI lifecycle, from data collection to model deployment to continuous performance testing.

Privacy by design/default and security by design/default, the concept of thinking through protecting personal and sensitive information during and not after the design phase, have been the gold standard for privacy, security, and data governance frameworks. These concepts still apply but now have to be expanded to accompany the often perpetual lifecycle stages and the unique challenges related to AI systems.

Organizations implementing AI, but also individuals using it should understand the characteristics of trustworthy AI, which include those outlined in the NIST AI Risk Management Framework[6] and the EU AI Act:

- Human-centric and human oversight
- Accountable and accurate
- Documented, transparent, and explainable
- Privacy-enhanced
- Valid and reliable
- Safe, secure, and resilient
- Fair with bias mitigated

Trust Is the New Currency

With data at the front and center of AI and much of today's data being personal or sensitive, trust is not a soft value; it's a strategic asset, an imperative value, and a mission statement. Organizations that operationalize AI governance to include privacy by default build trust with customers, regulators, and partners. Those that don't risk reputational damage, legal penalties, and loss of market share.

Between data breach awareness and growing AI literacy, AI users are becoming more protective of their personal information. As a result, those organizations building trustworthy and human-centric AI systems that focus on data given with full consent will win the race for acquiring the highest quality data with the lowest risk to drive their models and create a competitive advantage.

Governing the Future, Now

AI not only disrupts and transforms our lives, but it also changes the way we think about information, privacy, and trust. Generative AI systems can be both the chicken AND the egg: Its equal need for large, high-quality datasets to learn and process, but also its ability to create new data that then can become in itself another input are redefining data ownership, accountability, and what personal data means.

To thrive in the AI universe, we must rethink privacy data and AI governance not as a constraint, but as a design principle, helping organizations and individuals build and use AI not just to comply, but to compete, ethically, securely, and sustainably. After all, behind every AI decision is a human experience and personal data, like Kenn's or yours, shaped by invisible machines. Making those systems visible, understandable, and accountable is not just a technical challenge. It's a moral imperative that can and should be solved for. Knowing how privacy and AI coexist will help you, your organization, and society to evolve from "black box AI" to "glass box AI," advocating for and insisting on systems that are transparent, explainable, and trustworthy.

Take Away: AI Privacy and Governance Checklist

- *Offer model cards*: These are like nutrition labels for AI. They help organizations and end users understand the risks and assumptions behind AI decisions.

- *Provide data lineage of inputs/training data*: Trace where data comes from, how it's transformed, and where it flows. This is essential for accountability, trust, and data quality, especially when AI decisions are challenged.

- *Understand AI systems*: Gain knowledge about how AI systems work and their implications.

- *Implement privacy by design/default*: Embed privacy considerations into the entire AI lifecycle.

- *Ensure transparency*: Communicate clearly about AI systems and data usage.

- *Define accountability*: Establish roles and responsibilities for AI outcomes.

- *Enhance security and red-team AI systems*: Implement robust protections against misuse and breaches.

- *Maintain human oversight*: Ensure humans remain involved in high-stakes decisions.

- *Monitor bias and discrimination*: Regularly test AI models to mitigate bias and ensure fairness.

- *Capture and honor consent*: Implement mechanisms to capture and honor user consent.

- *Stay informed on regulations*: Keep up to date with global and local AI regulations.

Glossary of Concepts Introduced by the Author

Dynamic Governance is an adaptive and agile framework for decision-making and oversight that evolves in response to, and in tandem with, technological advancements, regulatory shifts, societal needs, and other external forces. It emphasizes continuous feedback, flexibility, and co-evolution between governance structures and the environments they operate within.

Shadow Data is inferred or imputed personal information generated by AI systems through the analysis of large, multi-modal datasets, such as socioeconomic, behavioral, or demographic data, without the individual's direct disclosure or consent. This data can include sensitive attributes like political views, sexual orientation, or immigration status, derived from circumstantial indicators (e.g., location, purchases, media consumption, voter registrations). Shadow data challenges traditional privacy safeguards, as individuals are often unaware of its existence and unable to exercise control over its use, raising significant ethical and regulatory concerns when used in decision-making processes.

References

1. Kashmir Hill, "Automakers Are Sharing Consumers' Driving Behavior with Insurance Companies," *New York Times*, March 11, 2024, https://www.nytimes.com/2024/03/11/technology/carmakers-driver-tracking-insurance.html (accessed July 13, 2025).

2. *EU Artificial Intelligence Act*, February 2, 2025, https://artificialintelligenceact.eu (accessed July 13, 2025).

3. *EU Artificial Intelligence Act*, February 2, 2025, art. 4, "AI Literacy," https://artificialintelligenceact.eu/article/4/ (accessed July 13, 2025).

4. *Smart Advertising Services—Global Strategic Business Report, Research and Markets*, July 2025, https://www.researchandmarkets.com/report/smart-advertising (accessed July 15, 2025).

5. Edelman Trust Institute, *2024 Edelman Trust Barometer: Global Report* (2024), https://www.edelman.com/sites/g/files/aatuss191/files/2024-02/2024%20Edelman%20Trust%20Barometer%20Global%20Report_FINAL.pdf (accessed July 13, 2025).

6. National Institute of Standards and Technology (NIST), *AI Risk Management Framework (AI RMF)*, January 26, 2023, https://www.nist.gov/itl/ai-risk-management-framework (accessed July 13, 2025).

About the Author

Bettina Lippisch is a technology strategist and transformative leader with over 20 years of experience at the forefront of digital innovation. Her cross-functional expertise spans compliance, operations, business strategy, product development, and marketing. She has successfully led major digital transformation, operational, and data strategy initiatives across a wide range of industries and company sizes.

Bettina's career is defined by her talent for simplifying complexity and guiding organizations through the rapidly evolving landscape of technology and now AI. Having lived and worked on two continents, she is passionate about building bridges between cultures, people, and technologies.

She holds certifications as an Artificial Intelligence Governance Professional (AIGP) and a certified Information PrivacyManager (CIPM), and she has earned two MBAs, one in technology management and another in global enterprise management. Bettina is a frequent speaker and contributor to conferences and has also appeared on industry podcasts, panels, and roundtables. Through her advisory and consulting work, she educates other leaders and senior executives on how to implement AI governance, privacy, and responsible data use and leverage it as a competitive advantage that drives faster innovation. Bettina serves as co-chair of the New York City-based German-American Scholarship Association, where she helps raise funds for transatlantic scholarships.

Outside of work, Bettina brings the same passion and curiosity to her personal life. She enjoys exploring new cultures through food and travel and is continually perfecting her culinary and bread-making skills.

Email: bettina.lippisch@digitaltransform.io
Website: https://www.digitaltransform.io
LinkedIn: www.linkedin.com/in/bettinalippisch

CHAPTER 18

YOU TOO, SHOULD BECOME AI-AUGMENTED—HUMAN-CENTERED AI FOR A SUSTAINABLE FUTURE

By Marion Løken, PhD
Digitalization Leader; Top 50 Women in Tech
Oslo, Norway

> *Do the best you can until you know better. Then when you know better, do better.*
>
> —Maya Angelou

Artificial intelligence is redefining our ways of working in general and at the office in particular. In many ways, AI is a threat; it poses risks of eroding worker privacy and autonomy, introducing or worsening bias, and creating challenges in transparency, explainability, and accountability, potentially amplifying existing issues. In addition, it can make a broad part of the workforce obsolete in the coming months. It can automate many of the tasks you were doing on a daily basis, and it can even do them better than you did. You are no longer competing with your peers—which was at least a fair competition—now, you're up against machines that operate 24/7, accomplishing in seconds what used to take you hours, with accuracy you only dreamed of. It is pointless to compete.

So, what's your next move? Fortunately, you—and I—don't have to become redundant. You can adapt, learn to collaborate with AI, and significantly enhance your abilities. You can become AI-augmented and, like the majority of the workers using AI, improve your job satisfaction. In this chapter, we'll explore how you can excel in your area of expertise by leveraging AI not just as an assistant to boost productivity, but as your trusted buddy. It's about you gaining that incredible processing power over all the data and best practices out there, while you provide AI with the crucial human understanding it needs. We'll also delve into how AI can level the playing field in ways that traditional DEI initiatives haven't been able to. Ultimately, we'll examine how a cybernetic teammate might be the missing piece of the diversity puzzle and how their inclusion can significantly boost both your individual and team performance and impact.

Ten Thousand Hours for Becoming an Expert

You probably already know that you can't be an expert in everything. If you've been working for a few years, you're likely developing strong skills in a specific area. However, you'll only ever be average or good at many other things. And, let's face it, some things you'll struggle with, requiring significant effort to improve. Expertise is the combination of deep knowledge and its effective application, refined to the point of intuitive judgment.

Malcolm Gladwell suggested it takes around 10,000 hours of dedicated practice to master complex skills. While the exact number of hours is debated, the core idea remains: Achieving expertise requires a substantial time investment and focused practice. The consequence is that you have to prioritize one area to excel at.

This is perfectly normal; after all, you're only human. Finding information, processing it, and practicing takes time. AI, however, can accelerate the path to expertise by curating, summarizing, and personalizing knowledge, helping you gradually build the depth of understanding and skill comparable to a PhD student—in a fraction of the time once required.

Focus on Your Strength

I have been working long enough to have gone from a workplace environment where you needed to be good at everything, and your performance review was mainly focused on your weaknesses—or, nicely put, your "improvement areas"—to a strengths-based approach where the thing people hired you for is the thing you are great at, as long as you are self-aware enough to know what you are not good at and keep those shortcomings—"developmental areas"—under control.

This shift has been a major positive change for the workforce. It's no surprise that it has led to higher employee engagement, greater productivity and performance, increased job satisfaction, and lower turnover rates. Everyone prefers to focus on their strengths and spend most of their time doing what they are good at. But this also raises the bar: To be considered truly good at your job today, you need to excel at something.

We have already seen how AI can accelerate the path to expertise. In addition, AI can help you further minimize the time spent on areas where you are less strong, either by guiding you toward just the essential skills needed to manage those weaknesses or by automating parts of the work altogether. In doing so, AI allows you to double down on your strengths and focus your energy where you can create the greatest impact.

10x Professionals

In tech, the people who are really great at something are a special kind of employee. They are the type that everyone wants to hire and would pay a lot for—the 10x developer. The 10x developer is an individual who is significantly more productive and impactful than an average one. What this means is that some developers are "10 times" more efficient than their average peers. It has nothing to do with the number of lines of code that they produce; it is more about the quality and efficiency of their outputs, whether those outputs are lines of code or something else.

This is true in tech, and it is true in other fields. Experts are just better at their jobs. They understand the problem to solve faster, they know what good or great looks like, and their solutions are more elegant and impactful. If you are not a 10x professional yet, you might want to become one. AI can be your accelerator, helping you close the gap to 10x performance faster because it helps you work smarter.

Workplace Inequalities and Diversity

Work life, whether you want it or not, is a competitive arena. It is highly codified and has historically favored individuals who fit the mold, behaving and performing as expected. DEI (diversity, equity, and inclusion) initiatives have attempted to level the playing field and bring more diversity to the office. However, the pressure to perform remains and will continue. Diversity has been found by McKinsey to be a promoter of financial performance, due to its ability to bring different perspectives to the table, making problem-solving broader, more complete, and less prone to bias. Despite this, individuals with diverse backgrounds still suffer prejudice, and their performance is often judged more harshly and severely than that of their more conformist peers.

AI offers the potential to mitigate these inequalities. AI, sometimes combined with robotic tools, can support individuals with diverse needs, overcoming neurological and physical barriers.

Furthermore, properly engineered AI can evaluate work objectively, eliminating human biases from reviews.

AI-Augmented Daily Work Life for All

Artificial intelligence presents an unmatched opportunity to empower everyone—regardless of background, neurological and physical ability—to excel in their roles and expand their scope, further leveling the playing field. Individuals who skillfully integrate AI into their work achieve significantly greater efficiency, deeper insights, and a more substantial overall impact. They evolve into 10x professionals. These individuals leverage AI tools not merely to accelerate what is expected of them tenfold, but rather augment their outputs.

Table 1: The Transformation of Tasks Through AI-Augmentation

Standard Tasks	Above-Average Expectation	AI-Augmented Capability
Routine tasks	Efficiently manage and minimize time on repetitive tasks through proficiency, optimization, prioritization, accuracy, and potentially delegation.	AI automates routine tasks by enabling their delegation to AI agents, freeing up valuable time for more complex strategic thinking and innovation.
Strategic decision-making	Make strategic decisions based on insights by analyzing available data and identifying relevant patterns to inform choices.	AI enhances strategic decisions by analyzing extensive datasets and identifying complex patterns with greater accuracy, leading to AI-assisted insights.

Communication	Communicate effectively by crafting clear and impactful messages tailored to their audience.	AI streamlines communication by assisting in the creation of tailored and precise messaging through the efficient instruction of AI assistants.
Experimentation	Experiment and iterate by testing new ideas and refining approaches based on results.	AI accelerates experimentation by enabling rapid prototyping and comprehensive testing powered by AI's analytical capabilities.

Furthermore, AI tools can provide professionals with access to in-depth, PhD-level information in domains beyond their direct expertise. This capability, driven by sophisticated "deep research" and advanced analytical thinking, "reasoning"—all powered by AI—fosters a more comprehensive understanding and facilitates innovative problem-solving.

I've built a long career working with data, statistics, and advanced analytics. I got my PhD in applied mathematics—analytical chemistry to be exact— and have been immersed in data for over two decades. I've always used data to extract information and tell stories that help people make better decisions. I'm considered an expert in my field—analytics, AI, and engineering—and even received the honor of being named one of the top 50 women working in tech in Norway in 2025. And let me tell you, before AI tools came along, this was definitely a skill that was hard to come by.

To put that in perspective, overall adult numeracy is relatively low. According to the OECD's 2023 Programme for the International Assessment of Adult Competencies (PIAAC), just 14% of adults in participating countries reached level four or five—the highest levels of numeracy. Only around 1% reached level five, which involves advanced mathematical reasoning, complex problem-solving, and interpreting data in unfamiliar or abstract contexts.

And it's not just about numeracy itself. There are also physical barriers to consider. People with visual impairment can't see graphical representations, and no text reader can truly make them "see" a graphic. Just imagine trying to understand a graph by having every single point read out to you. The mental load would be immense.

Anyway, the good news is that AI is changing all of this. AI can now assist with numerical proficiency, extract the important information from visuals, and craft a compelling narrative, just as well as I ever could.

The truth is, AI is leveling the playing field so fast that my hard-earned "edge" is disappearing. And ironically, I'm one of the ones building the tools to accelerate this transition. My expert knowledge is becoming accessible to everyone.

Accessible to All

AI tools are increasingly common in workplaces, and AI features are being integrated into everyday applications. While some AI tools had a competitive edge at the end of 2024, most mainstream options now offer similar capabilities. If your employer doesn't provide access, personal licenses are often surprisingly affordable, costing less than a t-shirt for the significant productivity gains they offer—at least for now.

From Productivity to Augmented Human Capabilities

We're moving from a focus on productivity to a focus on augmented human capabilities. The development of these systems is driven by the emergence of human-centered artificial intelligence (HCAI). HCAI prioritizes human needs, values, and capabilities in the design, development, and deployment of AI systems, rather than just technological advancements and productivity. The key idea, in line with UNESCO's Recommendation on the Ethics of AI, is that AI should augment human abilities and well-being. This means the tools you use should enhance your skills, creativity, and decision-making.

Your New Cybernetic Teammate

AI is not just another productivity tool. AI is now capable of complex tasks like reasoning, critical thinking, and problem-solving. AI systems are increasingly demonstrating abilities that approach or challenge the Turing test, raising discussions about the potential for artificial general intelligence (AGI), at least in the way we had defined it in the past.

The most straightforward way to start working with AI is to think of it as your assistant. You can delegate those tasks that you know how to do really well but find incredibly boring—the stuff you've always wished you could just skip. Typical examples include organizing ideas into a table, correcting your grammar, checking the tone of your emails, summarizing text, or making sure a document has all the required fields.

But let's be real, this is a pretty limited use of AI. It'll probably only boost your productivity by around 10% because these kinds of tasks are just a small part of what you do to get things done.

The next level is to integrate AI more deeply into your workflow. You and your AI tool become a team, it being a kind of "cybernetic buddy," bouncing ideas off each other and going back and forth to refine your common output.

This should help you produce better and faster results in your area of expertise. Then, you can start using AI to expand your capabilities into other areas. This means that, gradually, you might need fewer people with very specialized skills. For instance, if you're a marketer who's great at strategy and you need to create an end user survey, you can easily prompt AI to help you with that. Or if you need an illustration for the survey, no problem, you can generate one with a three-line command. You'd usually have to ask a product designer to do those things, but now you can handle it yourself. So, does this mean you don't need input from other humans anymore?

Nothing Beats a Human Team Empowered by AI

If everyone becomes pretty good at everything thanks to AI, do we still need to collaborate in teams with other people? The good news is yes. A team at Harvard Business School looked into this and found that teams, whether they used AI or not, were significantly more likely to come up with top-tier solutions. This suggests that there's something about human teams working together that goes beyond just yourself using AI. And, of course, teams that *did* use AI were even more likely to produce those really excellent solutions.

As more and more of us start using AI in our daily work and become AI-augmented, the next natural step is to think of AI not just as our personal helper or buddy for one-on-one tasks and interactions, but as a full-fledged member of the team. This is definitely possible from a technology standpoint, but are you ready, in the middle of a team discussion, to pause and say, "Hey, let's get our AI teammate's opinion on this"?

As we get better at including an AI teammate, it makes you wonder: Could AI be a new form of diversity—just not a human one? Like any form of diversity, it's a minority perspective, and it needs to be given space and be listened to if it's going to have a positive impact. McKinsey already found that companies that are early and extensive AI adopters outperform their peers in revenue growth and profitability.

The Human Advantage

You've probably heard that AI is even starting to outperform humans in areas we thought were uniquely human, like empathy. This raises the big question: Is there any human advantage left? Can a machine really replace everything you do?

Here's the thing: Both AI's limitations and its power come from the same source—context. AI can be more accurate than doctors in making diagnoses because it has been trained on a massive amount of data, including countless cases and all the best practices in the field. However, in the corporate world, while AI has access to best practices, it often lacks the full context of your specific company and the particular decision you need to make. The people who make the best decisions are usually those who have access to the most relevant information—or a lot of information and are skilled at picking out what's important. AI-augmented humans excel at decision-making because they use AI to process more information and then filter out what's relevant. In any effective human-AI collaboration, the human's role is to provide context to help the AI recommendations to be relevant.

Remember, AI operates based on statistics, generating the most likely output given the context it receives. I often use the "I love…" sentence as an example. If you prompt AI to finish that sentence without providing any context, it will give you the most frequently used phrase from its training data (a huge collection of literature, forum posts, etc.), which is likely to be "I love you." But if you change the context to, say, sports, you'll get "I love football/soccer" (since it's the most popular sport globally). If you narrow that context to India, you'll get "I love cricket." And if you specify a user group like "5-year-old girl," you might get "I love to dance." This is stereotypical, but it is what is most likely given the limited context.

So, to bring it back to the human advantage and how you can really excel and leverage AI: Your human superpower is providing AI with the most relevant context to enable the best possible decisions. This is how you and your teammates become AI-augmented, enhancing your skills and impact in a way that goes beyond simple productivity gains.

And part of that context is things like ethics, values, and all those intangible factors that we can't easily quantify with data. Sometimes, you still need human "intuition," which, in a way, is just data in disguise that a machine can't process.　　Therefore, wherever you and your colleagues are on your AI journey, you can use AI collaboration to accelerate your path to expertise, bypassing the 10,000-hour rule, and achieve 10x professional impact. As you and your colleagues embrace this shift, you can start thinking about the inclusion of an AI teammate to your team. Ultimately, the future of work isn't about humans *versus* AI, but about humans *and* AI, working together to achieve more than either could alone.

About the Author

Marion Løken is a renowned leader in data science, AI, and digitalization, holding a PhD and an engineering degree. Recognized as one of the Top 50 Tech Women in Norway in 2025, she specializes in using technology to drive business growth and address complex challenges. With a strong commitment to ethical AI and data usage, Marion promotes empowerment, inclusivity, and collaboration. As a public speaker, she inspires diverse audiences, advocating for sustainable, human-centered tech-driven solutions.

Email: marion.loken@outlook.com

LinkedIn: https://www.linkedin.com/in/marionloken/

CHAPTER 19

LEARNING IN THE AGE OF AI: WHAT MACHINES CAN'T TEACH US

By Colin Mansell
CEO and Founder, Skills U
Singapore

The illiterate of the 21st century will not be those who cannot read and write, but those who cannot learn, unlearn, and relearn.
—Alvin Toffler

Redefining How We Learn

For centuries, education has largely followed a familiar model: the transmission of knowledge from teacher to student, most often within the four walls of a classroom. At its heart, this approach has focused on an industrial age model, dependent on rote memorization of facts and procedures. But as we step further into the digital age, it becomes increasingly important to re-examine what skills and knowledge will matter the most now and in the future, as well as how we learn as people, and how technology will reshape the learning process in this brave new world.

To understand the impacts, it is important to start by recognizing that there are distinct types of knowledge, and that we acquire and master these in distinctly different ways:

- *Declarative knowledge* (knowing *that*): factual information, concepts, and ideas - such as the capital of France, or Newton's laws of physics.

- *Procedural knowledge* (knowing *how*): skills, processes, and techniques - like playing the piano, writing a persuasive essay, or debugging a software program.

- *Conditional knowledge* (knowing *when* and *why*): understanding the context in which to apply declarative and procedural knowledge - for instance, when to speak up in a meeting or why a particular business strategy will work in one situation but not another.

In the past, knowledge was centralized and hard-won, living in the minds of a privileged few. You would have to memorize information through rote repetition, and retain it for when it was needed. But in the digital world today, knowledge has been decentralized and commodified, and is available to all. Just think, anyone with a smartphone and an internet connection now has access to more information than the President of the United States did just a few decades ago.

Knowledge in the Digital Age

Declarative knowledge has perhaps been most dramatically reshaped by advances in digital tools. In a world where ChatGPT, Google, Wikipedia and YouTube are at our fingertips, we have radically optimized the way people access factual knowledge. Apps like Duolingo replace the flashcards of old, and use spaced repetition to systematically embed vocabulary into our long-term memory.

Meanwhile, platforms like YouTube offer high-quality, dual-coded learning experiences (combining video and audio) that are often far more engaging than many traditional classroom experiences. Channels like Veritasium or CrashCourse have arguably become more effective science teachers than the average high school classroom.

Yet even this is just the beginning. We're nearing a moment when personalized, AI-enhanced 'adaptive learning' becomes the norm. Imagine a world where a child who loves basketball but finds their science teacher dull, is taught Newtonian physics by an avatar of Michael Jordan, speaking in his voice, using real basketball examples to explain the laws of motion. All of the technology exists today to make this happen: generative AI and video, with voice synthesis and real time personalization - this reality is actually much closer than we realize.

In the near future, anyone will be able to generate customized learning content at the click of a button, fully tailored to their personal interests, learning style, and cultural context. This will make declarative knowledge vastly more accessible and engaging, reducing disparities and boosting global literacy and access to opportunity at scale.

For procedural and conditional knowledge, however, things are a little different. The reason for this is that people learn *how* to do things and *when* and *why* in a much more human way, something that is often missed when people think about how learning is changing in a digital world. Ironically, this means that the future of learning may in fact need to be more humanized, rather than further digitized.

Vygotsky and the Social Emotional Nature of Learning

Long before the rise of the internet, AI and digital platforms, psychologist Lev Vygotsky offered us a powerful insight: as humans we are fundamentally social-emotional beings looking for meaning, and learning is not a solitary pursuit, but a social-emotional process, driven by deeply personal drivers. His social learning theory emphasized that cognitive development happens through interaction

- through dialogue, collaboration, and shared problem-solving. We learn best not in isolation, but in relationship with other people.

At the heart of this model is the recognition that motivation and the "why" behind learning is the true driving force. A child learning maths or French is less driven by a desire for knowledge, and more by a need for connection and recognition, whether from parents, teachers, or peers. This extrinsic motivation - the desire to connect with others, to be accepted, recognized, or praised - plays a powerful role not only in childhood, but throughout our adult lives as well.

Equally important and potentially more powerful is our intrinsic motivation: the internal, personal drive to grow and self-actualize. When a learner can connect what they're learning to a meaningful goal, such as earning an MBA to increase their income to support their family, this clarity of purpose becomes a powerful source of momentum. When the "why" is clear and strong, the commitment to follow through grows stronger too, and it is often our closest friends, a coach, or a good manager who helps us to uncover what that is, and to push through the challenges that come with learning.

Another key concept that Vygotsky focused on, is what he called the Zone of Proximal Development (ZPD) - the space between what a learner can do independently and what they can achieve with the guidance of a more knowledgeable other. This could be a teacher, coach, or even a peer. Learning, he argued, is most effective in this zone, where support and challenge are balanced. As the learner gains confidence and competence, the support can be gradually withdrawn - a process known as scaffolding.

In today's context, the ZPD has even greater relevance. While information is now abundant, learning and personal transformation still require a fundamentally human element of empathy, understanding and support, especially in procedural and contextual learning - when developing skills, it is more often about building confidence, breaking through imposter syndrome, or navigating complexity. Vygotsky's theories reinforce the critical importance of human mentorship, coaching, and social and peer learning, which are often entirely missing when people think about digital learning systems.

My Personal Learning Journey

My own journey through the world of technology and learning has taught me one fundamental truth: Technology will cause massive changes to our society, but it is only as powerful as the people who use it, and people will remain at the centre of everything. Technology will undoubtedly have profound impacts throughout society and will replace many hundreds of millions of jobs in the years to come. But for those who can keep up, it also offers powerful new opportunities to create and thrive.

My career began in London's music industry in the early 2000s, witnessing firsthand how the internet was upending traditional business models. I saw entire floors of staff laid off at the record companies we worked for, as people shifted from buying CDs to downloading MP3s. This disruption was sobering, but it also revealed a powerful truth: new platforms like MySpace and YouTube were empowering creators to build their own communities, bypassing old gatekeepers entirely. The digital shift wasn't coming; it was already here.

When I turned 30, I moved to Vancouver and started a digital agency that created websites and apps for our customers. While coding, software tools, and the systems we developed mattered, our success really depended on the successful development of our people. Hiring, onboarding, managing and developing people - these were not technical problems, they were deeply human ones.

With that realization, I went on to start a technology school dedicated to helping young people enter the workforce with not only the technical skills they need to succeed in their careers, but also the mindset and soft skills to thrive, teaching students to think like meaningful contributors to the world: as builders, collaborators and problem-solvers.

I now work with an amazing global team of educators and technologists helping leading universities, employers and government agencies to design and deliver scalable reskilling programs, giving us a front-row seat to the global transformation that is happening in learning. We've built an AI powered career development platform that provides a suite of learning and career development tools, and while the technology is very impressive, what we've seen is that the most effective learning environments aren't built around lectures and exams. They're built around people: peers, mentors, coaches, and facilitators.

In the digital age, we are often led into believing that there is an app or a new AI tool that will fix everything. Actually it's the opposite: humans remain the key ingredient, and learning needs to recognise that, and to put people at the heart of every learning journey.

Hard Skills vs. Human Skills

A key distinction to consider is between hard skills (such as learning JavaScript or learning a language) and human skills (such as communication or leadership). These two different domains also require quite different approaches and environments in terms of how we develop these different types of skills.

Hard skills are often technological, and are generally procedural in nature. They are learned through practice, experimentation, and hands-on application. Critically, they are also most effectively acquired using Vygotsky's social-emotional learning model. Despite the promise of online learning from online platforms, purely online learning completion rates remain dismally low, often less than 3%. Why is this? Because learning to code, to design, or to play the piano requires human support, recognition and encouragement, from human experts and a community of peers. Without the social-emotional scaffolding of mentorship, peer learning, and accountability, most learners simply disengage.

Human or soft skills are perhaps more difficult and the most important to learn, especially because they cannot be replicated fully by machines. These skills include leadership, communication, collaboration, and emotional intelligence. They are not mastered through rote content consumption, rather they are gained through lived experience, deep personal reflection, and feedback from knowledgeable other people. Learning to lead a team or to build trust in a conversation requires self-awareness and a willingness to commit to personal change. It requires metacognition, which essentially means watching ourselves think, and a level of vulnerability that requires maturity and can often be deeply uncomfortable. Coaching, mentoring, and facilitated dialogue therefore become the primary tools for developing these abilities.

Although the hard skills, especially in technology, are becoming essential and will give those who learn them an advantage, it is the soft skills that will have an outsized impact on someone's career success. In a world of increasing automation and AI, it is the human skills that become the enduring differentiators.

The Genesis of Skills U

These insights and experiences culminated in the founding of the Skills U platform. We started with a core belief that skills (both hard skills and the human ones) have become the currency in the digital age. Our focus is to build and scale learning using AI powered technology, and yet to do so using a highly human-centric, social learning model. We see small agile technology teams now building billion-dollar companies, and yet it is not because of legacy advantages, but because of the way their people have learned to adopt technology, and more importantly, how well they are able to work together.

We recognize the gap: while technology has made content more accessible than ever, it has not made learning more effective. If anything, it has made it less human, and the soft skills that people so often say need to be developed, are being neglected. Skills U was designed to address that gap: to build a platform that deeply understands every individual, building a detailed profile on every learner through skills assessments, 360 reviews, identifying your motivations, career goals, values, and interests, to help you uncover your own "why", and to build highly personalized lifelong learning pathways for every learner.

The platform is highly modular and can be deployed into any learning community, whether it's a university, a company or a community of learners. Skills U provides a centralized platform that connects each learner with a personalized pathway that can include courses, workshops and career opportunities, as well as coaches and mentors who bring the essential human element back into the centre of the learning process. The powerful Skills Match engine, analyzes the gaps between where each learner is today, and where they want to get to, providing a truly personalized, adaptive and skills-driven pathway to help each learner succeed in their chosen career.

At our core, our mission is to make learning and development accessible to anyone, anywhere, on any topic. We are reimagining learning as lifelong, adaptive, personalized and humanized, and through the power of social learning. All of us already have access to all the learning content we could possibly imagine; we believe the real impact is realized when every individual has access to complete and human-centered support that makes learning effective: real people who believe in you, challenge you, and help you grow.

The Human Skills We Must Now Develop at Scale

As AI continues to evolve, it will work as an extension to us, to handle the cognitive heavy lifting - from recall to summarization to low-level analysis. The role of humans will therefore shift to what machines cannot do well:

- *Emotional intelligence*: Empathy, self-awareness, collaboration, and the ability to navigate interpersonal dynamics. As automation increases, these "soft" skills become the hardest to replace and the most valuable.

- *Creative and entrepreneurial thinking*: The capacity to generate ideas, experiment, pivot, and create value in ambiguous environments. Innovation is, by nature, a deeply human act that connects us to our core purpose.

- *Critical and systems thinking*: Moving beyond surface-level answers to understand deeper structures, implications, and trade-offs. This includes ethical reasoning and navigating complexity.

- *Resilience and self-management*: Managing one's emotions, staying focused under pressure, and sustaining motivation over long learning journeys, especially when outcomes are uncertain.

- *Communication and influence*: Whether in writing, speaking, visual storytelling, or motivating others into creating the future, the ability to convey ideas clearly and inspire action remains a uniquely human superpower.

- *Meta-learning*: The ability to learn how to learn. This includes setting learning goals, managing one's own progress, and adapting learning strategies. It is foundational in an environment where the content is ever-changing.

- *Ethical and civic literacy*: As AI systems shape our institutions, we must ensure that humans retain the ability to ask deeper moral questions, hold systems accountable, and build a fairer society.

A Shift in Learning Models

To support the development of these skills, we must shift from traditional models of education to dynamic, experience-rich, socially embedded models of learning. This includes:

- *Human-centered*: Providing personalized coaching, mentorship and guidance and emotional support.

- *Adaptive and personalized*: Meeting every learner where they are, aligning with their motivations and career goals, and designing aligned learning plans at scale.

- *Peer-based and social*: Developing peer-based learning communities and encouraging shared exploration, feedback and reflection.

- *Project-based learning*: Learners apply knowledge to solve real-world problems and by learning through doing, and learning in the flow of work.

- *Blended and experiential formats*: Using technology not just to deliver content, but to simulate real-world contexts and practice environments.

- *Life-integrated, lifelong learning*: Recognizing that learning happens everywhere - in the workplace, in families, in communities - and designing systems that support it across a lifetime.

Who Wins, Who Loses

In this future, winners and losers will not be determined by IQ, test scores, or degrees, but by mindset, adaptability, and access to high-quality learning experiences. Winners will include:

- Individuals who embrace lifelong learning and skill reinvention

- Organizations that prioritize learning culture and invest in human capability

- Coaches, mentors, and facilitators who help others navigate complexity

- Learners with access to personalized, contextualized, tech-enabled tools

Losers will include:

- Institutions that cling to outdated models of credentialism
- Workers who fail to reskill in the face of role displacement

- Organizations that treat learning as a cost center rather than a growth engine
- Communities without equitable access to digital infrastructure and support

Future Learning Architectures

To support this shift in learning, we must also reimagine the architecture of education itself. The learning systems of tomorrow will be:

- *Modular:* Allowing people to stack and mix credentials based on evolving needs.
- *On-demand:* Accessible anywhere, at any time, tailored to individual goals.
- *AI-augmented:* Where every learner has a personalized copilot to recommend resources, give feedback, and guide their journey.
- *Human-driven:* With skilled coaches and mentors facilitating transformation, not just knowledge transfer.

The promise of these architectures is not just more efficiency, but more equity: a world where opportunity is unbound by geography, privilege, or prior access.

Identity and Belonging in Learning

Learning is not just about acquiring skills; it's about motivation and perhaps even more so, it is about our identity. The most powerful learning journeys involve a shift in self-perception: from outsider to insider, from passive recipient to active contributor. This is why belonging, cultural relevance, and emotional safety are so central to effective learning design. Whether it's a young woman in Nairobi learning to code, a mid-career professional in Singapore exploring a new career in product management, or a frontline worker in Houston upskilling in digital tools to adapt to industry changes, the question isn't just "What am I learning?" but it's "Who am I becoming?"

A Call to Reimagine

We are entering a transformative era, where the tools at our disposal are more powerful than ever, and so too are the challenges we face. The convergence of AI, automation, and global connectivity demands not just smarter machines, but wiser, more adaptable humans.

We must reimagine learning as a profoundly human endeavor, one that integrates technology without losing the relational, emotional, and contextual elements that drive true personal growth. We must invest not only in content delivery but in the scaffolding of human support: coaching, mentoring, feedback, and community.

This is the future we are building at Skills U: a next generation platform where AI helps us to better understand ourselves, to map out our potential career paths, and to design truly personalized learning plans for learners everywhere. Not as a luxury, but as the new baseline for what education must become.

Let us not simply prepare people for jobs that may disappear. Let us empower people to become learners, leaders, and creators in a world of constant change. Not to compete with machines but to become the drivers of change. The real future of learning therefore lies in cultivating the qualities that define us: our curiosity, our empathy, our resilience, and our capacity to imagine a better future. As machines grow smarter, our challenge is not to mimic them, but to deepen the very traits they cannot replicate: our humanity, our judgment, and our capacity for meaning and connection.

About the Author

Colin Mansell is the CEO and Founder of Skills U, a global skills development platform that collaborates with leading universities, employers and government agencies to design and deliver large-scale upskilling initiatives. A passionate advocate for digital innovation and lifelong learning, Colin has founded and scaled multiple technology and education ventures and currently serves on several boards as an active investor in the future of work and education. Based in Singapore with his wife and two children, he is committed to creating economic opportunity and empowering individuals to thrive in the digital economy.

Email: colin@skillsu.com

Website: www.skillsu.com

LinkedIn: https://www.linkedin.com/in/colinmansell

CHAPTER 20

BLUEPRINT FOR TOMORROW: DIGITAL STRATEGY FOR FINANCIAL SERVICES

By Nithin Mathews, PhD
AI Researcher, Financial Industry Advisor, Startup Mentor
Zurich, Switzerland

> *The future is digital, decentralized, and permissionless.*
> *The question is not whether it will come, but whether you will be*
> *part of building it.*
>
> —Andreas Antonopoulos

Building the future of finance starts with recognizing the people we are building it for—both the tech-savvy, borderless digital natives and those still left out of the system, the unbanked. This chapter puts a spotlight on these two groups at opposite ends of today's financial system as they seem to represent both the biggest opportunities and challenges. It then examines where much of the industry's current investment is flowing (hint: into artificial intelligence and the compute infrastructure it requires) and asks whether that focus aligns with what these users actually need. Next, we introduce Bitcoin, a decentralized financial network, protocol, and asset tackling problems traditional finance has struggled with. Finally, we compare these two massive forces—AI and Bitcoin—to highlight their differences, common ground, and how they might converge to shape the future of financial services.

From a Borderless, Digital Generation to the Unbanked Populations

In major cities worldwide, a new generation of digital natives is leading a lifestyle that transcends national borders and jurisdictional limitations. They seamlessly use the internet across languages and cultures to live their lives, and they expect their finances to be just as instantaneous and global. For this borderless generation, banking isn't tied to a place or a 9-to-5 schedule—it's an always-on service ought to be accessible from anywhere and integrated into their digital lives. If a bank or payment app can't provide real-time, around-the-clock access with instant international transfers and up-to-the-second updates, these users will quickly switch to one that can. What used to be perks—like 24/7 chat support or same-day cross-border payments—are now basic requirements. Old banking limitations (waiting days for a transfer or only operating during business hours) feel to them like relics of a bygone era.

Meanwhile, a starkly different reality persists for a huge number of people. In many developing regions, over a billion adults remain unbanked—without access to basic bank accounts or formal financial services. They often live cash-only, relying on informal methods to save or borrow. Imagine having to hide your savings at home or rely on an expensive local moneylender because the nearest bank is too far away. The causes of exclusion vary: lack of nearby banks or internet, missing IDs and paperwork, political instability, or deep mistrust of institutions after years of corruption and failures. For these communities, something as simple as a safe place to save money or an affordable loan can be life-changing, yet it remains out of reach.

There are signs of hope. In the past decade, innovative solutions have started to chip away at the problem. Mobile money services like M-Pesa in Kenya or bKash in Bangladesh show that basic financial services can reach people through a device almost everyone has: a mobile phone. These services let users store money digitally, send payments via text, and even get small loans, all without ever visiting a bank. Technology like this allows countries to leapfrog brick-and-mortar banking by delivering financial access over mobile networks. But technology alone isn't enough. True inclusion also requires better infrastructure, supportive policies, and education.

In short, apart from typical clients, two distinct groups are putting significant pressure on today's financial system: one of hyper-connected customers pushing finance to be instant and borderless, and another of vast unbanked populations for whom even basic digital banking is still a dream. The future of financial services must bridge these worlds by creating truly global, always-on services while extending basic access to those left behind. In practice, the industry has to innovate with cutting-edge tech *and* build out infrastructure and trust where they're lacking.

Where Is the Current Focus of the Financial Services Industry?

So, where is the financial industry focusing its efforts today? Recent trends show that much of the new investment is pouring into technology, especially artificial intelligence (AI) and the massive data centers behind it. Huge funding is going into AI research, AI chips, AI foundational models, and compute infrastructure as countries and corporations race for AI leadership. This promises impressive advances. Imagine banking systems that catch fraud the instant it happens, or AI advisors that give you personalized financial tips on demand or even anticipates your needs. However, there's a concern: If all this innovation serves mainly the already connected and tech-savvy, will it help those who lack even basic banking? The worry is that progress could create a new divide, where a "digital elite" enjoys AI-powered finance while poorer, disconnected communities fall further behind.

AI in finance began by boosting efficiency behind the scenes—automating paperwork, speeding up loan processing, and flagging fraud in real time. Now it's moving to the frontlines with intelligent "AI agents" enhancing customer experience. These advanced AIs can analyze data, make decisions, and interact with users autonomously. Picture an AI that optimizes your finances 24/7 or chats with you about your spending. Banks and fintechs are already rolling out AI-driven investment advisors and smart budgeting apps to serve more people at lower cost. Many experts predict that soon many routine financial decisions—approving loans, detecting suspicious transactions, offering budgeting advice—will be handled largely by AI in the background. This could make banking faster, more personalized, and potentially fairer by removing some human bias.

But these same AI tools can also be turned to fraud and manipulation. Criminals are also using AI to create convincing scams: fake voices to authorize illicit transfers, phony emails and messages that look exactly like they are from your bank, even deepfake videos to spread false news and sway markets. In short, it's getting harder to know what's genuine. Financial security now requires new defenses to verify identity and information in a world where AI can fabricate very realistic fakes.

These concerns have fueled a push for responsible AI. Financial institutions and governments alike are emphasizing that AI must be transparent, fair, and accountable. Principles like avoiding bias and explaining AI decisions are starting to be codified. The EU's proposed AI Act, for example, is one of the first major efforts to set "rules of the road" for AI, and other countries are issuing their own guidelines and regulations. Essentially, there's a worldwide effort to guide AI development, so we get the benefits without the disasters.

Another challenge is authenticity—knowing what's real when AI can fake so much. Old CAPTCHA tests to distinguish humans from bots are failing. New ideas include digital watermarks embedded in AI-generated content and using cryptography to create tamper-proof records of data or decisions. The aim is to maintain trust: If an AI makes a decision about your money, or a sensational video goes viral, there should be a reliable way to verify its origin and legitimacy.

In summary, AI is opening exciting possibilities in finance, but it's also forcing a rethink of security, privacy, and trust. The industry is right to invest in AI's potential, but it must equally invest in making AI trustworthy.

A Universally Accessible, Decentralized Financial Network

While big banks and tech giants pour resources into AI, a very different financial revolution has been growing from the grassroots. In 2009, amid a global financial crisis, an anonymous developer, known as Satoshi Nakamoto, introduced Bitcoin—a radically new kind of money. Unlike traditional currencies issued by governments or digital payments run through banks, Bitcoin runs on a decentralized network of users and doesn't rely on any central authority. It showed that people could send value to each other online without a bank in the middle, a revolutionary idea both at the time and even more so today.

Bitcoin is a decentralized network, a protocol, and a digital asset (i.e., coins). Bitcoin solves the problem of digital scarcity. It uses cryptography and a public ledger to ensure that its coins cannot be duplicated or double-spent. Every transaction is recorded across the network, and no single party can alter the records or create fake coins. This lets strangers anywhere trade value over the internet with confidence, without needing a trusted intermediary to verify transactions.

The implications are huge. For the digital native, Bitcoin offers a financial system as global and always-open as the internet itself. You can send money worldwide anytime, and it arrives in minutes, with no bank needed. For the unbanked person, all that's required to join the digital economy is a basic phone and an internet connection; you don't have to wait for a bank to reach your village. There have even been cases of humanitarian aid delivered via Bitcoin straight to those in need, bypassing corrupt middlemen. In essence, Bitcoin created a parallel financial network: If the traditional system isn't accessible or efficient enough, the Bitcoin network is an alternative that anyone online can use.

A key difference is Bitcoin's monetary policy. Unlike central banks that can print currency at will (sometimes sparking inflation), Bitcoin has a hard cap of 21 million coins, and new coins are released on a fixed schedule that slows over time. No one can arbitrarily change this. Thanks to this predictable scarcity, Bitcoin is often likened to digital gold—a finite store of value. Many people have indeed turned to Bitcoin as a hedge against inflation or as protection against unstable local currencies.

By 2025, Bitcoin has edged closer to the mainstream. Some companies hold it as part of their reserves, and regulators have approved Bitcoin-based investment funds, which fold it into everyday portfolios. Major financial institutions now offer Bitcoin-based services and products to their clients. What began as an experiment is increasingly treated as a legitimate asset class and ledger—perhaps even a building block for new financial products—rather than just an internet oddity.

Two Trillion-Dollar Industries Shaping the Future of Finance

The AI revolution is mostly driven by centralized entities in rather top-down fashion, while Bitcoin's network is decentralized and bottom-up as it was adopted initially be cryptographers, computer scientists, and retail users and only finding its way into institutions in the recent years. That is, while AI services are built and controlled by large companies or governments, Bitcoin is run by a global community with no single entity in charge. Using AI-driven platforms generally means trusting the provider. You give your data and transactions to a company's cloud (or any kind of infrastructure) and trust their algorithms—

whereas using Bitcoin requires no such trust in a middleman (the motto is "don't trust, verify," since anyone can check the public ledger themselves). AI platforms often require permission and accounts (you have to sign up and abide by their rules), while Bitcoin is permissionless and open to anyone with an internet connection. AI's infrastructure is concentrated in giant data centers that handle enormous volumes of transactions and data, whereas Bitcoin runs on many distributed nodes and prioritizes security and transparency over sheer speed. And when it comes to change and control, AI systems can be quickly updated or reconfigured by their owners or regulators, but Bitcoin's rules change only very slowly and require broad consensus, making it more rigid but also more resistant to sudden shifts.

Despite these differences, AI and Bitcoin can also be complementary to each other. AI excels at analyzing data and automating complex tasks, often within controlled networks. Bitcoin provides a neutral, tamper-proof transaction layer that doesn't depend on any one organization's control. Together, AI could handle intelligent decision-making, while Bitcoin ensures the underlying transactions and records are open and reliable.

Convergence and Future Potential

One surprising convergence of AI and Bitcoin is around energy. AI data centers need huge, steady amounts of electricity, while Bitcoin mining is also energy-intensive but uniquely flexible. In some places, Bitcoin miners are helping balance the electric grid by using excess energy when supply is high and powering down when demand peaks. For example, miners can soak up surplus power from wind or solar farms at off-peak times, then shut off during a consumption spike so that homes and businesses have enough. This buffer role helps prevent wasted energy and can even encourage more renewable projects, something traditional always-on data centers can't easily offer.

Another intersection might be in the agentic future our digital world seems to be heading into. Consider a future where AI agents developed by different banks, fintech companies, or even clients across the globe communicate and transact directly with one another. For instance, an AI assistant from a U.S. retail bank might coordinate with an AI from a Japanese e-commerce platform to instantly settle a cross-border purchase for a customer, or a corporate treasury AI in Germany could negotiate short-term liquidity arrangements with an AI at a Singaporean investment firm—all in real time and without human intervention. In these scenarios, the question of what currency these autonomous AIs should use becomes critical. Traditional national currencies are tied to specific jurisdictions and subject to varying regulations, clearing delays, and exchange risks. Centralized digital currencies or stablecoins, meanwhile, are still issued and controlled by companies or governments, which introduces counterparty risk and potential for censorship. Bitcoin, by contrast, stands out as the most natural and the only internet-native option for value exchange between such global AI agents. It operates independently of any single country or corporation, is accessible worldwide, and settles transactions according to open, transparent rules that anyone can verify. Since Bitcoin does not belong to any one entity and can be used in a permission-less manner by any AI agent with an internet connection, it provides a level playing field for autonomous value transfer. This makes it uniquely suited as the universal digital currency for machine-to-machine payments across borders, industries, and regulatory regimes—a truly global medium of exchange for the age of intelligent automation.

Meanwhile, both technologies are becoming mainstream and coming under greater oversight. Bitcoin has gained legitimacy with clearer rules and growing institutional adoption. AI has become widespread enough that governments around the world are crafting guidelines to manage it. There's a broad acknowledgment that these innovations need to be steered responsibly as they integrate into everyday life.

Looking ahead, how might AI and Bitcoin work together to improve finance? On the optimistic side, we could get the best of both. AI would make financial services smarter, more personalized, and ultra-convenient, while Bitcoin makes them open, global, inclusive, decentralized, permission-less, and transparent. You might have an

AI advisor managing your finances 24/7 and executing transactions via the Bitcoin network instantly and globally. Geographic borders and bank hours would become irrelevant. Even someone in a remote village could access cutting-edge financial tools through such AI-and-Bitcoin powered platforms, enjoying the same opportunities as someone in a major financial center.

Making this future work for everyone means building inclusivity, transparency, and accountability into the system from day one. Developers should craft AI and Bitcoin tools that are user friendly and fair. For example, AI models that decide on loans or insurance should be explainable and checked for bias, and Bitcoin onramps for newcomers should be secure and simple to use. Policymakers will need to cooperate internationally, since these technologies ignore national borders. We might need global standards for things like digital identity verification, data privacy, and ethical AI use in finance to ensure these tools aren't used to exploit consumers or evade laws.

There are early moves toward this inclusive approach. Some are calling for certain digital services to be treated as public infrastructure—open and accessible to all. India's UPI platform, for instance, made digital payments easy and almost free for millions of people, acting as a public utility for money transfers. If such platforms integrate advanced AI for smarter services and connect with the open network that is Bitcoin, they could help ensure that even underserved communities benefit from high-tech finance. And because Bitcoin is open-source, the technology remains in the public realm, where it can be scrutinized and improved by a broad community rather than controlled by any single entity.

In conclusion, fusing AI and decentralized finance is a unique opportunity to reinvent financial services. We have the tools to make finance far more accessible, efficient, and user friendly. We can imagine a world where sending money or getting a loan is as easy as sending a text, and where an AI looks out for your financial health. But achieving that vision requires intentional effort to include everyone. We must continually ask: Who benefits from this innovation, and who might be left out or hurt? Then we need to adjust our designs and policies to close those gaps.

If we get it right, the future of financial services could be one of shared prosperity where we onboard the next billion people to our financial world. A future where technology truly breaks down barriers instead of reinforcing them. It's an ambitious goal, and reaching it will demand creativity, collaboration, and vigilance—but the reward is a fairer, more inclusive financial world that serves the entire humankind rather than a selected few.

About the Author

Nithin Mathews was born in India and raised in Switzerland. He completed his studies in computer science in Switzerland and Germany, and holds a PhD in artificial intelligence, robotics, and decentralized systems from the Université Libre de Bruxelles, Belgium. With a decade-long career in academic research, Nithin has published numerous articles in scientific journals, including *Nature Communications*, garnering over 2,000 citations according to Google Scholar. Several of his publications have also attracted significant public interest, receiving coverage from prominent media outlets such as TechCrunch, Engadget, *The Wall Street Journal*, and the Discovery Channel.

Currently, Nithin is an Industry Advisor at Microsoft, serving within their Worldwide Financial Services Industry team. In this role, he advises Microsoft's strategic clients in the financial services sector, helping them shape technology-driven business strategies. Additionally, he mentors startups and emerging businesses. As part of his separate mandate with Innosuisse, the Swiss Innovation Agency, Nithin evaluates and provides expert guidance on science-based projects, fostering pioneering research and promoting industry-academic collaborations to drive innovation within Switzerland.

The views and opinions expressed in this chapter are solely those of the author and do not necessarily reflect the official policy or position of Innosuisse, Microsoft, or any of its affiliates. In the spirit of practice informing theory, Nithin confirms that he applied generative AI tools for the research, structuring, and editing of this chapter.

LinkedIn: https://www.linkedin.com/in/nithinmathews/

CHAPTER 21

THE GRID THAT THINKS

By Patrick J. Meyers
Energy Executive—LC Energy Partners; Author: *Tariff Nation*
Houston, Texas

The most powerful energy source in the 21st century isn't solar or wind. It's coordination.

—*Jesse Jenkins, energy systems engineer and Princeton professor*

The EV That Saved the Grid

At precisely 4:46 p.m. on a sweltering Thursday afternoon in July, the lights in Houston flickered—but they didn't go out. That morning, the Texas grid operator had issued a level two emergency alert. Record demand, caused by relentless triple-digit heat, had outpaced expectations. It was the kind of day when the grid typically walks a tightrope between brownouts and blackouts, relying on a dusty peaker plant to fire up in desperation. Only this time, help didn't come from a hulking gas-fired turbine on the edge of town. It came from a driveway.

Specifically, the driveway of a retired firefighter named Raul Alvarez, who, unbeknownst to him, had just become a power plant operator. His Ford F-150 Lightning, plugged into his garage outlet, sent 15 kilowatts of electricity surging back into the grid—enough to offset six homes' air conditioning for a crucial 30-minute window. He didn't touch a button. He didn't even know it happened until he got a $47 credit from a company he'd barely heard of—one of the rising players in a new business called "virtual power plants." What made Raul's truck more than a vehicle—what turned it into infrastructure—wasn't the lithium battery. It was an algorithm.

From Central Command to Digital Swarm

For the last hundred years, power flowed like authority in a Cold War bureaucracy: top-down. Big centralized power plants—coal, gas, nuclear—pushed electrons out across high-voltage transmission lines to cities, towns, factories, and homes. Engineers in control rooms called the shots, flipping switches and ramping turbines in a carefully choreographed sequence that kept the lights on. The job of the customer? Pay the bill and stay out of the way. That system worked beautifully—until it didn't.

Solar panels began popping up on rooftops like mushrooms after a spring rain. Batteries followed. Smart thermostats. EVs. Soon, households weren't just passive consumers—they were microgenerators, storage units, and flexible loads. The grid, once a symphony conducted by a single baton, turned into a jazz improvisation session with millions of players.

The only way to make that kind of system work is with something smarter than any human conductor—something that can make decisions in milliseconds, ingest petabytes of data, and orchestrate an invisible ballet of electrons across time zones, weather patterns, and price spikes. That something is AI.

This is where the virtual power plant—or VPP—steps in. It's not a place. It's a platform. A decentralized digital utility stitched together from thermostats, inverters, electric vehicles, home batteries, and industrial HVAC systems, all coordinated by software. It doesn't own any of these devices. It just tells them when to turn on, off, up, or down. It's Uber, but for energy. Airbnb, but for electrons. And its most powerful employee? An algorithm that never sleeps.

Where utilities once managed a few dozen giant machines, today's VPPs are learning to manage a few million small ones. That's not just a change in scale—it's a revolution in architecture. And at the center of it is AI, transforming chaos into coordination.

The Brain Behind the Swarm

If the traditional power grid was a Roman aqueduct—massive, hierarchical, and designed to flow one way—then the AI behind a virtual power plant is more like the neural network of an ant colony. It doesn't just react—it anticipates. It learns. And it does it at the edge of chaos, where old-school control rooms would've thrown up their hands.

Imagine a hot summer day. Power prices are spiking. A cloud bank is rolling in over San Diego, threatening solar output. Meanwhile, a fleet of Teslas in Phoenix is reaching peak charge demand just as the local substation is approaching its capacity limit. A human operator might notice one of those things—after it's too late. The AI sees them all, and more. In microseconds.

That's because a modern DERMS—distributed energy resource management system—can crunch variables from across the grid:

- Local weather radar
- Wholesale power market prices
- Battery state-of-charge across 50,000 rooftops
- EV charge curves
- Forecasted air conditioning demand
- Transmission congestion
- Even your Nest thermostat setting at 2:15 p.m.

Then it makes a decision: Discharge six megawatts from batteries in Sacramento, pre-cool 4,000 homes in Phoenix by two degrees, and delay EV charging in Fresno by 15 minutes. No one notices. But the grid stabilizes.

These systems don't run on rules—they run on reinforcement learning. Like chess engines, they simulate millions of outcomes, constantly updating strategies based on feedback. They "learn" how to win the game of grid stability. And the game isn't just about reliability anymore—it's about economics.

The AI isn't just trying to keep the lights on. It's trying to maximize value for everyone in the swarm. It knows when to charge batteries with low-cost solar at noon, and when to sell it back at $250 per megawatt-hour during a 6 p.m. crunch. It arbitrages in real time—matching physics to finance, electrons to incentives. And most people will never know it happened. That's the brilliance of it.

The Virtual Utility in Action

Let's rewind to Germany—2021. A sleepy town outside Munich. The residents of a new suburban development signed up for a program with *sonnen*, a German battery manufacturer turned energy platform company. For a few euros a month, the homeowners agreed to let their batteries be "networked" into something called the *sonnenCommunity*.

When demand surged, the AI would dispatch their batteries. When solar spiked, it would soak up excess. If prices crashed, it would charge. When prices soared, it would sell. Every device was automated, optimized, and coordinated—by software in the cloud.

One day, a transmission line fault knocked out part of the local grid. The utility didn't respond fast enough. Sonnen's AI did. Within seconds, it islanded the community into a self-sustaining microgrid, powered by solar, batteries, and careful load control. Not one homeowner lost power. A nearby block went dark. This one didn't flicker. To the residents, it felt like magic. In reality, it was a distributed, AI-coordinated ballet of electrons, synchronized in real time by something no bigger than a laptop.

Meanwhile, across the Atlantic, Tesla was quietly assembling one of the largest virtual power plants in the United States—in California. By networking thousands of Powerwall home batteries, Tesla's AI dispatched power back to the grid during emergencies, earning grid credits for homeowners and offsetting peak demand.

But Tesla didn't invent the VPP model. That honor goes to a patchwork of upstarts like AutoGrid, Sol-Rite, and EnergyHub. These firms realized early that power no longer had to come from a plant—it could come from anywhere, so long as someone (or something) was smart enough to control it. And that someone was never going to be a human.

Meet a Few of the Architects of the Virtual Grid

It's tempting to think the power grid of the future will be built by the same utility giants that built the last one. It won't. It's being reimagined by software companies—ones you've probably never heard of—who are stitching together millions of devices into what amounts to a ghost utility. No smoke stacks. No wires. Just code. Let's meet three of them.

AutoGrid—The Algorithm That Ate the Grid

If there's a "brain" quietly running the distributed grid revolution, it might be AutoGrid. Headquartered in Silicon Valley, AutoGrid doesn't generate electricity. It doesn't sell solar panels. What it sells is control—specifically, algorithmic control of everything from your EV charger to a 100-megawatt battery field. Its flagship platform, AutoGrid Flex, is a kind of global conductor for distributed energy: It ingests millions of data points in real time—weather, market prices, load forecasts, customer usage—and turns those into actionable instructions. Charge this battery. Curtail that load. Dispatch those EVs. All in milliseconds.

The company has already orchestrated over 6,000 megawatts of flexible capacity across 17 countries, quietly working behind the scenes for major utilities like NextEra, National Grid, and

TotalEnergies. In other words, AutoGrid isn't replacing the power plant—it's replacing the person who decides when to turn it on.

SOLRITE—*When Your Garage Becomes a Power Plant*

In San Marcos, California, a quiet energy revolution is playing out in the most unassuming of places: the American garage. SOLRITE doesn't build power plants. They build contracts. Their offer is elegant in its simplicity: zero upfront cost solar panels and two sonnen batteries—40 kilowatt-hour of clean energy storage—in exchange for a 25-year fixed-term escalating rate power agreement. No maintenance. No hassle. And—here's the kicker—they become part of a virtual power plant.

SOLRITE is wiring together ordinary rooftops into an invisible machine—a living, breathing swarm of decentralized energy. Thousands of homes that once passively consumed electricity now stand ready to push power back to the grid, stabilize frequency, ride out blackouts, and soak up excess solar. It's as if someone took the concept of Airbnb, strapped it to a lithium battery, and taught it how to read the power markets.

This isn't a utility. It's a network. One that stretches across Texas, California, and Puerto Rico—not with steel and concrete, but with code and coordination. What the 20th century built with substations and smokestacks, SOLRITE is rebuilding with algorithms and rooftops. The future grid won't be centrally owned. It'll be locally optimized—and quietly humming in your garage.

EnergyHub —*The Smart Home Whisperer*

While AutoGrid thinks big and SOLRITE installs the batteries, EnergyHub plays where you live: inside your smart thermostat, your EV charger, your HVAC unit. If there's a company that's quietly turned residential demand response into an art form, it's this one. EnergyHub's platform links millions of smart devices to utility partners, enabling precise control during moments of peak demand.

When the grid gets tight, EnergyHub can nudge 50,000 thermostats down two degrees, slow the charging of a few thousand EVs, and delay a laundry cycle—enough to flatten a peak without anyone noticing. That kind of flexibility used to require a gas peaker plant. Now it just takes a signal from the cloud.

Their model is pure orchestration. EnergyHub doesn't manufacture devices. It doesn't install anything. It simply connects and controls—and it's working with over 60 utilities across North America to deliver reliable, real-time capacity with no new infrastructure at all. If SOLRITE is the developer and AutoGrid is the brain, EnergyHub is the grid's behavioral psychologist, changing usage patterns one smart device at a time.

The Big Idea: Software as a Substation

Together, these companies are rewriting the rules of the energy economy:

- AutoGrid makes chaos manageable.
- SOLRITE and *sonnen* make batteries profitable.
- EnergyHub makes behavior grid-compatible.

None of them owns a single generator. And yet, they're bidding capacity into wholesale markets, winning utility contracts, and stabilizing the grid more effectively than many legacy players. They represent a shift from infrastructure to intelligence, from hardware to optimization, from monopoly control to algorithmic markets. If you're looking for the next Exxon or Duke Energy, don't look for smokestacks. Look for source code. Because the future grid won't be built. It'll be trained.

Arbitrage, Resilience, and the New Middlemen

There's a quiet revolution underway in the energy markets, and it's being led not by engineers in hard hats, but by quants with laptops. In this new world, the most valuable asset isn't a gigawatt-scale turbine—it's a well-timed decision. And AI, unlike any utility executive, can make thousands of them per second.

Suddenly, a 13-year-old girl in Bakersfield whose home battery discharged four kilowatts into the California grid during a price spike is, functionally, a participant in a wholesale power auction. Not because she understands locational marginal pricing, but because her VPP software does.

This is energy arbitrage at utility scale. It buys low and sells high—not just across time zones, but across technologies. It knows when to soak up wind power in Oklahoma, store it in a neighborhood battery in Tulsa, and sell it to the grid during the evening peak in North Texas. And all of it happens invisibly—like high-frequency trading, only with electricity.

In fact, it's already begun to look a lot like the evolution of finance. Power is no longer just generated and consumed—it's brokered. Aggregators are emerging as the new middlemen of energy: firms that don't generate electrons themselves but use AI to control, package, and monetize the electrons flowing from homes, vehicles, and commercial buildings.

If you squint, these aggregators begin to resemble the early days of Uber or Airbnb. No infrastructure, no physical product. Just control. Just code. And just like those platforms, they're redefining what ownership means. In the future, your car might earn more as a battery than as a mode of transportation.

The Tension: Regulation, Trust, and the Fear of Black Mirror Grids

But not everyone is thrilled with this techno-utopian vision. Ask a grid operator in the Midwest what keeps them up at night, and they won't say terrorism or storms. They'll say "algorithms we don't control." The reality is that VPPs—especially when controlled by private companies—pose a very real challenge to legacy utilities, regulators, and even basic public trust.

Who decides when your fridge shuts off for a five-minute frequency response? Who owns the data from your smart inverter? What happens when 10,000 EVs in Los Angeles all start charging at once because an AI made a bad call?

There's also the matter of regulatory lag—policy frameworks built for coal plants and nuclear reactors, now trying to govern a system where your air conditioner is bidding into a market. In some states, VPPs still aren't even legally recognized. In others, they're hamstrung by rules that treat distributed energy like a nuisance instead of a resource.

And then there's the looming *Black Mirror* scenario: AI-driven power systems that, in the wrong hands or with the wrong training data, spiral into chaos. A misfire in logic, a bug in an update, a hacked API—and you don't just lose your internet connection. You lose your grid. As one utility executive bluntly put it: "We built the grid for predictability. AI gives us performance—but with mystery."

AI Didn't Just Optimize the Grid. It Democratized It.

What we're watching is the grid's Napster moment. The end of top-down control. The birth of a networked, decentralized, democratized, AI-optimized power system. And like all revolutions, this one didn't ask for permission.

AI didn't just make the grid smarter. It reshaped its economics, its physics, and its politics. It took power—literal and figurative—from centralized monopolies and redistributed it to homes, neighborhoods, and devices, all whispering to each other through the cloud.

If the last century of energy was about scale, this one is about orchestration. And orchestration, in a world of infinite complexity, belongs to the machines that can handle it. The real story here isn't about the software or the hardware. It's about the shift in mindset: from consumers to participants, from power plants to platforms, from energy as a commodity to energy as an intelligent service. The next utility won't own power plants. It won't even own wires. It will own algorithms. And for the first time in history, the grid won't just respond to us. It will listen to us. And if we're lucky, it'll learn.

About the Author

Patrick J. Meyers is a seasoned energy executive and strategist with over two decades of experience at the intersection of power markets, technology, and infrastructure. He has advised leading firms in renewable energy, AI-driven grid solutions, and utility-scale battery deployment, with a focus on helping companies navigate the transition from centralized generation to distributed, intelligent systems. Patrick is the founder of LC Energy Partners and has held leadership roles spanning business development, market strategy, and M&A across the US and Latin America.

Patrick holds a bachelor of science degree in engineering from the University of California, Los Angeles (UCLA), and an MBA in analytical finance and econometrics from the University of Chicago Booth School of Business. When he's not guiding energy firms through transformation, he's writing about the future of power, trade, and resilience.

Patrick is the author of: *Tariff Nation – How to Navigate and Prosper in the Trade Wars Ahead.* https://www.amazon.com/dp/B0FLX12TR9

Email: Patrick.meyers@tariffnation.us

LinkedIn: https://www.linkedin.com/in/patrick-meyers-114346/

CHAPTER 22

FROM TECHNOLOGY TO TRANSFORMATION: AI AS A CEO'S STRATEGIC LEVER

By Ángel Moyano
AI Transformation Strategist
Madrid, Spain

The only true voyage of discovery would be not to visit strange lands but to possess other eyes.

—Marcel Proust

When the Data Was Right, but the System Wasn't Ready

A few years ago, I was advising one of the world's largest energy companies. They had just rolled out a powerful new AI algorithm designed to monitor oil wells and predict when production might start to decline. It used real-time data and clever algorithms to catch early warning signs—things that even experienced engineers might miss.

On one of their key sites, the system sent up a red flag: A high-performing well was showing subtle signs that something was off. The AI recommended an early inspection to avoid a costly breakdown. The digital team quickly passed the insight to operations.

But nothing happened. The field team didn't ignore the alert—they just didn't have a process for acting on it. The AI wasn't officially part of how maintenance decisions were made. There was no budget set aside to follow up on an algorithm. So, the alert was acknowledged… and quietly set aside.

Three weeks later, the well started to fail. Production dropped, equipment had to be shut down, and emergency resources were scrambled. The AI had been right, but the organization wasn't ready to respond.

That moment changed everything. The company didn't just improve the technology; they changed how they worked. They made sure AI insights were included in weekly planning meetings. They created clear roles for people to act on those insights. And they built trust between the digital team and the operations crew.

What made the difference? Not the software, but the structure. Because even in a company managing billions in assets, the real obstacle wasn't the machine. It was how decisions were made.

That experience didn't just challenge my assumptions. It reframed them. It became clear that the true power of AI wasn't in the technology itself, but in how organizations are designed to act on it. This realization underpins the central thesis of this chapter: To unlock real value, we must treat AI not as a digital tool, but as a strategic lever embedded at the heart of how organizations operate.

AI as a Strategic Imperative: A Wake-Up Call for Leaders

AI is no longer a futuristic buzzword. It's a present-day imperative. Across industries, organizations are facing an inflection point. Competitive dynamics are shifting rapidly, driven by a convergence of economic, technological, regulatory, and social pressures. The companies that treat AI as a strategic catalyst—not just a digital tool—are pulling ahead. Those that delay risk irrelevance.

In conversations with C-suite executives across the globe, one question continues to rise in urgency: "How can we stay competitive in a world moving at the speed of AI?" Too often, the responses focus on technology pilots, automation use cases, or innovation labs— tactical reactions to a strategic challenge. These surface-level efforts rarely move the needle because they fail to address the deeper issue: AI is not a project. It's a paradigm shift.

As someone who has spent over two decades leading digital transformations for global enterprises, I've seen this mistake repeated across industries. Organizations remain locked in pilot purgatory— experimenting at the margins without reimagining the core. What's missing is not ambition, but alignment. AI must be treated as a leadership agenda and embedded into the fabric of the operating model. The urgency is real. The following five interconnected forces are reshaping the context in which leaders must now operate:

1. *Rising Complexity Across the Enterprise*

From new regulations to evolving customer expectations, organizations must navigate an increasingly complex ecosystem. AI helps cut through this complexity, enabling faster data-driven decisions in environments of uncertainty and risk.

2. *Overlapping Transformations*

Digital, operational, and ESG (environmental, social, and governance) shifts are converging, creating simultaneous and often conflicting transformation demands. AI provides the connective tissue to prioritize initiatives, track ROI in real time, and orchestrate execution across silos.

3. *Capital Efficiency Under Pressure*

With rising capital costs and shorter investment horizons, stakeholders demand faster returns with lower risk. AI enhances capital allocation by improving forecasting accuracy, reducing uncertainty, and optimizing deployment across strategic initiatives.

4. *Erosion of Stakeholder Trust*

Investors, customers, and employees now expect visibility, purpose, and proof of long-term relevance. AI can support transparency by linking performance to predictive insights, strengthening strategic narratives, and reinforcing accountability.

5. *Market Volatility and the Need for Agility*

From supply chain shocks to geopolitical upheavals, volatility is no longer episodic; it's constant. Traditional planning models are too slow. AI enables real-time scenario planning and adaptive responses that help firms stay ahead of disruption rather than merely reacting to it.

These forces are not temporary; they're structural. They demand a new leadership mindset, not just new tools. The organizations that thrive will be those that rethink how value is created, how decisions are made, and how talent is empowered. That is the true AI advantage: not just pilots, automation, or analytics, but a reinvention of the enterprise operating model—led from the top. Anything less is a missed opportunity.

Beyond Pilots and Hype: A Reality Check

It's easy to get swept up in the excitement of AI. Executive teams approve innovation budgets, data scientists build dazzling proofs of concept, and early results from pilot programs spark enthusiasm. But then... the momentum stalls. The pilots don't scale. The models don't get embedded. The excitement fades. This story is more common than many leaders would like to admit.

According to a 2024 McKinsey Global Survey, 88% of companies report some level of AI adoption, yet only 5% have successfully transformed a full business domain to deliver consistent, measurable impact.[1] Similarly, a recent BCG study found that 74% of firms struggle to scale the value of AI across their organizations.[2] These statistics reveal a growing disconnect between ambition and execution—between deploying AI tools and realizing strategic transformation.

So, why does this happen? Too often, AI is delegated downward—confined to innovation labs, emerging tech teams, or tucked inside digital transformation offices with limited authority. These groups are encouraged to "experiment," "move fast," and "show what's possible." And they often do. But without organizational alignment, their work rarely moves beyond the prototype stage.

Instead of integrating AI into core business rhythms, many companies keep it at the edges, creating a system where insights exist, but action doesn't follow. The enterprise is wired to protect the status quo, not to translate new capabilities into new ways of working. It's not a technology problem; it's a leadership one.

The deeper issue is structural. Organizations treat AI like software—something that can be deployed, measured, and iterated on. But AI is not plug-and-play. It changes the nature of decisions, roles, incentives, and workflows. It reshapes the business, not just enhances it. That's why a new operating model is essential, one built for intelligence, not just automation. In this model:

- AI is embedded into the way decisions are made, not bolted on as a reporting layer.
- Cross-functional teams lead development and deployment, breaking down silos between tech, business, and operations.
- Governance isn't about control—it's about clarity. Who owns the outcome? Who can act on the insight? Who ensures it aligns with strategy and ethics?

- And most critically, leadership doesn't just supervise the change. It sponsors, steers, and sustains it.

Consider a company that deploys an AI model to optimize pricing. It might show that dynamic pricing increases margin, but if commercial leaders aren't prepared to change contracts, sales tactics, or commission structures, the insight remains academic. This is the AI adoption trap: results on paper, resistance in practice.

To move beyond pilots and hype, organizations must treat AI not as an initiative, but as a shift in how the business operates. That requires rewiring roles, retraining talent, restructuring incentives, and redefining what good leadership looks like in an AI-driven world.

Architecting an AI-Centered Operating Model

Once organizations acknowledge that AI isn't a side project but a fundamental shift in how value is created, the next challenge becomes operational: How do we build a business that can actually run on intelligence? This requires more than inserting algorithms into workflows. It calls for a deliberate redesign of the operating model—how people make decisions, how teams collaborate, how performance is measured, and how resources are deployed. In an AI-first world, the business itself must become a dynamic system capable of learning, adapting, and scaling intelligence across every function.

Let me be clear: AI does not simply enhance existing structures; it pressures organizations to evolve. A compelling example comes from a global retailer that successfully embedded AI across its supply chain. By connecting demand forecasting with pricing engines and inventory optimization, they improved forecast accuracy by 20% and reduced working capital by 25%. Impressive numbers, but the true unlock wasn't the technology. It was the operating model behind it: a centralized AI platform, empowered cross-functional teams, and strong executive sponsorship that ensured insight translated into coordinated action.

From my experience advising global firms, three design elements consistently separate organizations that scale AI from those that stall:

1. *Executive-Led Governance*

AI must be owned at the top. When decisions about data, models, and risk management are siloed under IT or innovation labs, strategic alignment is lost. Boards and C-level executives must treat AI as a business agenda, not a technical one. This means:

- Setting bold AI ambitions linked to enterprise goals
- Owning the ethical and regulatory frameworks
- Ensuring funding, talent, and incentives are aligned

Governance isn't about controlling the model—it's about directing the value.

2. *Cross-Functional Delivery at Scale*

AI success depends on integrated teams, not isolated expertise. The most effective organizations deploy cross-functional squads that include data scientists, business owners, designers, and engineers working together in agile sprints. This structure:

- Accelerates feedback loops and iteration
- Ensures business context informs technical decisions
- Builds shared ownership for outcomes, not outputs

In AI-native organizations, value isn't generated in silos. It's orchestrated across disciplines.

3. *The Role of AI Translators*

AI translators—or what some call "bilinguals"—are the often-overlooked glue that makes AI real. These professionals can speak the language of business and the language of data science. They don't write the code, but they know what the code needs to solve. They don't own the P&L, but they know how to tie technical outputs to commercial impact. Translators are essential for:

- Framing the right business problems
- Defining success metrics that matter
- Driving adoption and trust across business lines

In organizations where AI delivers sustained value, these roles are institutionalized—not improvised.

Putting It All Together

An AI-centered operating model is not a blueprint. It's a living system. It requires ongoing calibration, cultural change, and leadership intent. But when built well, it enables a new kind of organization: one that can sense changes in the environment, learn faster than competitors, and mobilize its people and technology to respond in real time. AI is not the heart of the business, but it's becoming its nervous system. To activate it fully, leaders must not just adopt AI, but architect for it.

AI as a Leadership Agenda

One of the most persistent and costly myths in enterprise strategy today is the belief that AI belongs to someone else, typically IT, innovation, or the data science team. But the truth is clear: AI is now a leadership responsibility, not a technical initiative.

AI doesn't just change how systems operate. It changes how organizations think, learn, and compete. That shift demands a new kind of leadership, one that blends strategic foresight, digital fluency, and human-centered judgment. Leaders must not only understand what AI can do, but also reimagine how it should shape business models, decision rights, performance metrics, and cultural values.

Take the role of the CIO, for example. Traditionally viewed as a steward of infrastructure and systems, the CIO must now operate as a business architect, designing intelligent operating models, enabling data-driven cultures, and acting as a bridge between enterprise ambition and digital execution.

But this shift extends beyond the CIO. In the age of AI, every executive—CHRO, CFO, CMO, COO—must reframe their function through a new lens:

- How can AI unlock better outcomes for our customers and employees?

- What workflows, incentives, or skillsets need to evolve?

- What ethical boundaries must we define and defend?

Leading with AI doesn't mean becoming a technologist. It means orchestrating change, ensuring the organization is structured, resourced, and culturally equipped to turn intelligence into impact. At its core, this leadership agenda includes:

- Championing purpose before platforms: AI should be deployed to advance business strategy—not for novelty's sake.

- Shaping a culture of continuous learning: Curiosity, not certainty, is the most valuable trait in an AI-driven enterprise.

- Embedding ethics and inclusion from the start: Responsible AI is not just a policy—it's a practice.

- Using AI to augment, not replace, human judgment: The goal isn't to eliminate people from the loop—it's to elevate their capacity for insight, empathy, and creativity.

In the right hands, AI doesn't replace leadership. It redefines what great leadership looks like. And at the center of that leadership is purpose—not just profit, not just productivity, but a commitment to responsible progress.

A Call to Action for Today's Leaders

We are standing at a profound inflection point. AI is not the destination—it's the instrument. The true journey is not about how advanced the technology becomes, but about how courageous and prepared we are to lead through it. This is not a future to delegate. It's a future to design.

As leaders, we must recognize that the real transformation isn't happening in data centers or algorithmic models—it's happening in boardrooms, planning sessions, and performance reviews. The decisions we make today will determine whether AI becomes a force for strategic reinvention or another missed opportunity filed under "innovation theater." To lead effectively in an AI-enabled world, we must:

- Elevate AI to the executive agenda: Not as a tech trend to monitor, but as a core lever of competitiveness, growth, and purpose.

- Rewire operating models for intelligence at scale: Move beyond isolated use cases and build enterprise systems where AI informs how decisions are made, how teams are structured, and how value is delivered.

- Invest in digital fluency and collaboration across the enterprise: Equip people at every level—not just with tools, but with the confidence and clarity to use them wisely.

- Embed ethics, empathy, and accountability into every AI initiative: Trust is not a given. It must be earned through transparency, alignment with values, and a commitment to human well-being.

The organizations that will thrive in this next era won't be those that simply deploy more algorithms. They'll be the ones that reimagine how they work, decide, and lead. They will treat AI not as an overlay or add-on, but as a foundational blueprint for reinvention.

This is our moment of leadership. Not to react to disruption, but to architect resilience. Not just to keep pace with technology, but to define what progress should look like. So let us not ask, "What can AI do?" Let us ask, "What kind of leaders must we become to use it wisely, boldly, and with purpose?" Because the future of AI is not waiting to be discovered. It is waiting to be led.

References

1. *The State of AI in 2024: Generative AI's Breakout Year*, McKinsey & Company, 2024, https://www.mckinsey.com/capabilities/quantumblack/our-insights/the-state-of-ai-in-2024-generative-ais-breakout-year (accessed July 2025).

2. *AI Adoption in 2024: 74% of Companies Struggle to Scale Value*, Boston Consulting Group, October 24, 2024, https://www.bcg.com/press/24october2024-ai-adoption-in-2024-74-of-companies-struggle-to-achieve-and-scale-value (accessed July 2025).

About the Author

Ángel Moyano is a senior digital strategist with over 20 years of experience leading high-impact technology and business change across global enterprises. Based in Madrid, Spain, Ángel has held leadership roles at both Boston Consulting Group and Accenture, where he has shaped and executed digital and AI strategies for some of the world's most complex organizations across sectors including energy, financial services, insurance, etc.

Currently a partner and director at BCG, Ángel specializes in designing future-ready operating models that embed AI, analytics, and digital capabilities at the heart of how organizations work and compete. His track record includes managing enterprise-wide transformations impacting thousands of employees, orchestrating portfolios of hundreds of initiatives, and generating over a billion euros in value for clients.

Ángel is also deeply engaged in the academic and professional development community. He serves as an adjunct professor at Instituto de Empresa and Universitat Oberta de Catalunya, where he teaches topics related to digital business strategy, IT systems, etc.

Ángel's passion lies in working with C-level leaders to move beyond technology hype—to redefine competitiveness through human-centered, purpose-driven innovation. His work bridges strategy and execution, ensuring that AI and digital tools are not just implemented, but embedded into the decision-making fabric of the enterprise.

LinkedIn: https://www.linkedin.com/in/angel-moyano-jimenez/

CHAPTER 23

BUILDING TRUST IN AN AGENT-DRIVEN WORLD

By Giorgio Natili
Head of Engineering
Seattle, Washington

AI is evolving faster than ever, but the organizations that will thrive in the agent-driven world aren't those that move the fastest; they're the ones that move most responsibly. In my experience building AI systems at scale, one truth has become clear: The companies that succeed are those that embrace what I call the "trust imperative": the belief that long-term innovation depends not just on capability but on credibility.

Unfortunately, the current narrative around AI leadership is dangerously narrow. Too many executives see AI primarily as a cost-cutting tool—a way to reduce headcount and automate away operational expenses. That mindset isn't just short-sighted; it's actively harmful. Leadership isn't about using AI to replace people. It's about preparing them for better, more meaningful roles.

To be clear: AI should enhance human capabilities, not replace them wholesale. We should absolutely use AI to eliminate unsafe tasks that are either dangerous or highly repetitive, especially those that put people at risk. But when it comes to knowledge workers, the opportunity lies in augmentation, not substitution. The most effective AI deployments I've seen create human-AI partnerships that amplify creativity, accelerate decision-making, and unlock outcomes neither humans nor machines could achieve alone.

When Speed Kills Trust

The chaotic 72-hour saga of Windsurf in July 2025 illustrates how acquisition velocity can shatter the trust that sustainable innovation demands. OpenAI's $3 billion acquisition offer, made in May 2025, expired in July. Within hours, Google DeepMind struck a $2.4 billion deal to hire Windsurf's CEO Varun Mohan, co-founder Douglas Chen, and key research leaders while also obtaining non-exclusive licensing rights. By the end of the weekend, Cognition acquired what remained: Windsurf's product, IP, and the 250 employees Google had left behind.

Sovereignty in Limbo: Who Owns the Code, the Data, and the Promises?

Windsurf's hundreds of thousands of users face a basic but unresolved question: Who is responsible for their code, data, and intellectual property? Windsurf's privacy policy covers standard acquisitions, but this split creates unprecedented governance challenges. While Cognition has promised that Windsurf will "continue to operate as they have been," they also plan to integrate the company's IP into Cognition's broader platform, introducing entirely new conditions for how user data might be processed, accessed, or retained. There were not the terms users originally agreed to.

The Broader Pattern

Windsurf may be an extreme case, but it's not an isolated one. Increasingly, we're seeing a pattern: acquiring the IP, hiring the talent separately, and leaving behind the human infrastructure. This treats AI development like physical manufacturing—as if you can buy the factory without losing anything essential. But AI development isn't just code or compute. It's an ecosystem of decisions, assumptions, and evolving relationships.

For users, this creates a new category of risk: sovereignty fragmentation. When the people who understand your requirements work for one company, another processes your data, and a third owns the licensing rights, where does accountability live? Who governs the system's behavior? These are not just legal questions, they're trust questions. The agent-driven economy, with its reliance on autonomous systems and distributed decision-making, demands models of growth and consolidation that preserve—rather than fragment—the trust relationships that make AI systems viable in the first place.

AI Agents: The Next Trust Challenge

The Windsurf story highlights a growing problem: As AI systems become more embedded, more autonomous, and more distributed, trust becomes harder to maintain. This shift isn't just about how AI is deployed. It's about what AI is becoming. We're entering the era of intelligent, autonomous systems: AI agents.

An AI agent is an autonomous software entity that can perceive its environment, make decisions, and take action to achieve specific goals without constant human intervention. Unlike traditional software, which follows fixed instructions, agents can adapt their behavior in response to new inputs, learn from experience, and interact dynamically with other agents or humans.

What distinguishes agents from conventional automation isn't just scale. It's autonomy. When systems act independently—integrating across ecosystems, handling sensitive data, and making high-impact decisions—the traditional trust model no longer applies. We can't rely solely on static controls or after-the-fact audits. We need systems that enforce trust at runtime—cryptographically, autonomously, and at scale.

Technical Infrastructure: The Operational Backbone of Trust

When autonomous agents act on sensitive data in real-world environments, trust isn't something you declare. It's something you prove. The second pillar of data sovereignty is about making that possible. It's about infrastructure that enforces rules at runtime, exposes tampering, and makes trust verifiable by design. This starts with a breakthrough that changes the terms of what's possible: confidential computing.

Confidential Computing: Trust in Execution

The most significant breakthrough in AI privacy isn't happening in policy. It's happening in silicon. Confidential computing represents a shift in how we approach data protection. It allows sensitive data to remain encrypted not just at rest and in transit, but also in use. This fundamentally changes what's possible for AI systems. Agents can now run secure, policy-governed AI workflows across cloud platforms without ever exposing sensitive data, not even to the cloud provider.

Confidential computing offers a cryptographically verifiable way to use public cloud resources while maintaining control. Your data can be processed without ever being decrypted outside of secure hardware. In short: you don't just control where your data goes; you control how it's processed and by whom.

Remote Attestation: Verifying the Invisible

Confidential computing protects the execution environment, but how do you prove that protection to others? That's where remote attestation comes in. This capability enables external parties to cryptographically verify that an AI agent is running in a genuine, uncompromised environment before they release sensitive data or authorize privileged actions.

When an agent starts in a confidential computing environment, the hardware generates a cryptographic "measurement" of everything loaded into the secure enclave—from the operating system to the application code. This is signed by the hardware itself, producing an attestation report that proves exactly what software is running and in what security state.

Attestation isn't a one-time event. It can be continuous, re-verifying the enclave's integrity after updates or changes. If the environment is compromised or unexpected code is loaded, trust is revoked and sensitive operations halt.

This trust model extends across the stack: Immutable container images guarantee software reproducibility, hardware security modules (HSMs) protect the cryptographic keys, and policy engines embedded inside the enclave ensure agents stay within governance constraints, even under adversarial conditions.

Data Control and Interoperability: Governing Without Isolation

The third pillar of data sovereignty—data control and interoperability—addresses perhaps the most complex challenge in the agent-driven economy: How do we maintain meaningful control over our data when that data must move, integrate, and be acted on by autonomous systems across boundaries.

Policy That Travels with Data

Traditional security models assume that control comes from containment: Build walls around your data. Restrict access and audit usage after the fact. But in an agentic environment, those assumptions collapse. Agents need to exchange information, invoke external services, and integrate across jurisdictions. The solution isn't stronger walls; it's portable, enforceable policy.

This is where policy-aware data becomes essential. Instead of simply encrypting data, we embed machine-readable usage policies into the data itself. When an AI agent accesses this data, it inherits both the content and the rules governing how that content can be used. Policy-aware data ensures that governance survives the journey across systems, agents, and borders.

Standards and Scaling

To make this work at scale, we need interoperable protocols and open infrastructure—tools that let agentic systems negotiate, comply, and coordinate without sacrificing control.

The Model Context Protocol: Standardizing Agent Negotiations

Originally designed to help applications pass context to large language models, model context protocol (MCP) is evolving into something deeper: a standard for agent-to-agent negotiation. Before exchanging data, agents can declare their constraints and capabilities. This negotiation happens before any sensitive data is exchanged. MCP ensures that agents establish terms of engagement upfront, building trust through verifiable intent.

LlamaStack: Open Infrastructure for Agentic Workflows

Meta's LlamaStack provides the complementary foundation: a fully open-source framework for building and operating agents at scale. Unlike closed systems that lock developers into proprietary infrastructure, LlamaStack gives organizations full control over how agents run, communicate, and enforce policy.

It also democratizes access. What once required highly specialized teams and heavyweight engineering can now be prototyped and deployed by smaller organizations without compromising on sovereignty, privacy, or composability. But with this democratization comes responsibility. As open-agent ecosystems grow, we need to ensure that standards and infrastructure evolve with privacy and governance built in from the start and not added later under pressure.

Transparency and Accountability: Making the Black Box Legible

Even the most advanced technical infrastructure is meaningless if we can't understand what our agents are doing or why they made the decisions they did. As AI systems gain autonomy and take on increasingly sensitive roles, transparency and accountability are no longer optional. They are essential conditions for scale, governance, and public trust.

Explainability: Beyond Technical Compliance

When I talk about explainable AI (XAI) in the context of autonomous agents, I'm not referring to technical outputs meant for data scientists. I'm talking about systems that can articulate their reasoning in terms that are intelligible and actionable for people affected by their decisions.

This level of clarity achieves three things: It helps humans understand the decision; it provides a path for remediation; and it creates an audit trail of the agent's reasoning and alignment with policy. That last point is critical: Explainability is not just about transparency; it's the gateway to enforceable accountability.

Immutable Audit Trails: Trust That Can't Be Erased

Transparency tells us what happened. Accountability requires proof and verification that nothing was hidden, altered, or quietly undone.

Every significant action taken by an autonomous agent, especially in high-stakes contexts, must be logged in a way that cannot be silently erased or manipulated. Basic logging isn't enough. Timestamps can be forged. Records can be selectively deleted. Trustworthy systems require cryptographic immutability.

But integrity alone isn't sufficient. The audit trail itself must be semantically rich. We need to capture not just what the agent did but also why it made that choice, what alternatives it considered, and what policies guided its decision. Only then can we enable both automated policy enforcement and human oversight at scale.

The Human-AI Partnership: Division of Labor

Transparency and accountability make agentic systems safer, but they don't make them wise. That's where humans come in. The most effective AI implementations treat humans and AI not as rivals but as complementary partners. Agents excel at high-throughput decision-making, pattern recognition, and consistency at scale. Humans excel at ethical reasoning, contextual nuance, and judgment under ambiguity.

This partnership model also reshapes how we think about work. Instead of replacing humans, agents free people to focus on higher-value activities. When agents are auditable and explainable, and when humans remain in the loop for what truly matters, we don't just scale efficiency. We scale trust.

Ethical Sentinels: Programming Values into Autonomous Systems

As AI agents become more autonomous, we face a deeper challenge: How do we ensure they act in accordance with human values, even when no human is watching? This isn't just a technical question. It's an ethical one. And the more decisions agents are empowered to make, the more urgently we need to answer it.

Beyond Rule-Following: Adaptive Ethical Reasoning

Most AI ethics frameworks still focus on rules. These are important foundations, but for truly autonomous systems, they're not enough. Agents operating in the real world won't just encounter clear-cut rules. They'll face conflicting values, ambiguous contexts, and edge cases with no single correct answer. That's why we need systems capable of adaptive ethical reasoning, systems that can understand not just the letter of the law but its spirit.

Cultural Sensitivity and Global Operations

Ethical reasoning doesn't happen in a vacuum. What's seen as responsible behavior in one region may be deeply problematic in another. As AI agents operate across borders, they'll need to navigate not just regulations but cultural expectations.

This isn't just about legal compliance. It's about cultural respect and human-centered design. It requires systems that learn from local context; feedback loops that enable adjustment; and teams with cultural fluency to anticipate before it happens. Diverse, globally minded teams aren't just a compliance checkbox; they're a strategic necessity for any organization deploying agents at scale.

Executive Responsibility: Embedding Ethics in Practice

This is where ethics becomes personal, because every executive deploying AI is already making ethical choices, whether they acknowledge it or not: Choosing to prioritize efficiency over transparency is an ethical choice. Choosing cost savings over capability enhancement is an ethical choice. Choosing to move fast and break things when real people bear the consequences is an ethical decision.

Ethical leadership means owning these trade-offs. It means defining clear values early; embedding them into technical systems, not just policies; and creating mechanisms that surface concerns before they become scandals. Most importantly, it means understanding that ethics is not the enemy of profit; it's a prerequisite for trust. And trust, in the age of agents, is the most valuable business asset of all.

Regulation as a Catalyst for Innovation

France's push for comprehensive AI governance at the EU level isn't just about protecting citizens. It's about creating conditions for sustainable innovation. By establishing clear, enforceable rules for AI development and deployment, the EU is building a regulatory environment in which companies can invest confidently, knowing the ground won't shift beneath them. That clarity matters enormously for agentic systems. When you're deploying AI agents that will operate autonomously, sometimes for months or years, you need confidence in the policy landscape.

For global companies, this creates a crucial strategic choice: Build separate systems for different regulatory jurisdictions, meeting only the bare minimum in each, or build to the highest standard from the outset and earn trust everywhere. Those that build to the EU standard from the beginning not only move faster in new markets, they often command premium pricing for systems that are verifiably trustworthy.

Data Sovereignty as Competitive Advantage

For organizations operating in business-to-business environments, the ability to prove that client data remains under control throughout the lifecycle of an AI system is becoming a decisive factor. That's where confidential computing becomes more than a privacy safeguard. It becomes a sales tool. Technologies that support cryptographic attestation—showing that sensitive data remains encrypted even during processing—let vendors offer something far more compelling than legal assurances: technical proof of governance.

The Road Ahead: Security Challenges That Demand Solutions

But the systems we've built so far are just the beginning. If we want to secure the next generation of agent-driven infrastructure, we'll need to solve challenges that today's best practices still can't address.

- *Model extraction and inversion attacks*: As AI agents become more valuable, attackers are developing sophisticated techniques to extract proprietary model parameters or reverse-engineer training data from deployed systems. We need defenses that protect intellectual property without compromising agent performance or transparency.

- *Multi-agent system security*: As agent ecosystems scale, dozens or even hundreds of agents may operate in parallel. We need new frameworks for modeling and securing distributed, autonomous systems, not just individual agents.

- *Adversarial manipulation in agent networks*: Not all agents will be trustworthy. We need robust detection mechanisms to identify manipulation, contain its spread, and safeguard network integrity.

- *Cross-domain policy enforcement*: Agent workflows increasingly span jurisdictions, organizations, and regulatory regimes. We need policy engines that reason across boundaries without losing granularity or control.

- *Quantum-resistant security*: Practical quantum computing will render many of today's cryptographic protections obsolete. We need to begin the transition to quantum-safe algorithms before the cost of delay becomes irreversible.

- *Real-time audit and compliance*: Most audit systems were built for periodic review, not real-time autonomy. We need a compliance infrastructure that can keep pace, verifying decisions and logging actions without slowing the system down.

Solving these problems won't just secure the future; it will define it. The future belongs to leaders who recognize that security isn't a constraint on innovation; it's the foundation that makes enduring innovation possible. But security is only part of the equation. The question isn't just how we build trustworthy agents. It's what kind of society we want them to serve. That's where the human dimension comes back into focus.

The Human Renaissance: Technology in Service of Human Flourishing

As we build increasingly sophisticated AI systems, it's easy to lose sight of the ultimate goal: using technology to enhance human potential, not replace it. The agent-driven economy isn't about creating a world where machines do everything while humans become obsolete. It's about creating a world where technology amplifies what makes us most human.

Cognitive Liberty in the Age of AI

One of the most profound challenges facing AI leaders is preserving cognitive liberty—the freedom to think and choose without manipulation. This isn't a distant concern. Recommendation algorithms already shape what information we see, what products we buy, and even whom we date. As AI agents grow more autonomous and more persuasive, their influence over human decision-making will only deepen.

Preserving cognitive liberty requires intentional design. It means creating agents that present meaningful choices rather than preordained outcomes. It means giving users visibility into how decisions are made and control over how their preferences are used. Above all, it means recognizing the true purpose of AI is to expand human freedom, not to constrain it.

Digital Safe Spaces: Confidential AI in Practice

This is where confidential AI becomes not just a technical breakthrough but a moral imperative. When people can engage with AI systems without fear that their private thoughts, vulnerabilities, or struggles will be exposed, we create the conditions for authentic expression and growth. Confidentiality isn't just about data protection; it's about psychological safety.

Confidential computing makes this possible. An AI assistant running in a trusted execution environment can provide sophisticated support while cryptographically ensuring that conversations remain private. It can learn, adapt, and improve without ever exposing the identity or content of individual sessions.

Preparing for Tomorrow: The Skills That Matter

As leaders shaping AI-driven organizations, we bear responsibility for helping people adapt, not just technologically, but humanistically. This doesn't mean turning everyone into a coder. It means investing in the skills that will matter most in an AI-augmented world: critical thinking, emotional intelligence, ethical judgment, and creative problem-solving. When AI handles the routine, it's the human ability to make meaning—to navigate ambiguity, to sense nuance, to act with empathy—that becomes irreplaceable.

Building the Future: A Call to Responsible Leadership

As we stand at the threshold of the agent-driven economy, the choices we make today will shape the digital world our children inherit. We can build AI systems that respect human dignity, preserve individual agency, and serve the common good. Or we can build systems that optimize for short-term profits while externalizing their costs to society.

The technology itself is neutral. The difference lies in leadership—in the values we embed; in the principles we refuse to compromise; and in the vision we hold not just for what AI *can* do, but for what it *should* do. I believe the future belongs to organizations that choose the harder path: building AI systems that are not just powerful but trustworthy; not just efficient but ethical; not just profitable but principled. These organizations will earn the trust that becomes the foundation of sustainable competitive advantage in an AI-driven world.

The agent-driven economy isn't inevitable in any particular form. It's a future we're actively constructing through the choices we make today. Let's build it to serve human flourishing, not just human productivity. Let's build it to expand human potential, not constrain it. Let's build it to honor the best of what technology can offer while preserving the best of what makes us human.

About the Author

Giorgio Natili is the head of engineering at a leading technology company in Seattle, where he focuses on building trustworthy AI systems at scale. He has been at the forefront of confidential computing implementations and serves as an advisor to organizations navigating the intersection of AI innovation and data sovereignty. His work spans the technical, ethical, and strategic challenges of deploying autonomous AI systems in highly regulated environments. Giorgio's vision and mission is to explore how artificial intelligence can reshape human-centered technology, focusing on its ethical opportunities to advance inclusion, strengthen trust, and protect digital dignity. His work spans the agentic economy, where autonomous systems redefine how we create and collaborate, and the pursuit of trustworthy AI, built on transparency, accountability, and fairness. He cares deeply about cognitive liberty in the age of AI—the right for people to think, decide, and act without undue technological influence.

As an engineering leader, Giorgio champions responsible AI leadership, building cultures rooted in authenticity, vulnerability, and trust, and helping teams thrive through transformation. At the core of his mission is a commitment to digital accessibility and inclusive design, ensuring that the technologies we build today empower everyone, not just a few.

LinkedIn: https://www.linkedin.com/in/giorgionatili/
Website: www.giorgionatili.ai

ORCHESTRATING TRANSFORMATION: STRATEGIC LEADERSHIP AT THE INTERSECTION OF AI, GOVERNANCE, AND CUSTOMER EXPERIENCE

By Mohamed Omer
AI Advocate
Frederick, Maryland

AI is not an assistant but an orchestral instrument—leaders must conduct with empathy, purpose, and ethical clarity.
—Kashif Zaman

The Threshold of a New Era

A CEO of a fast-growing fintech company once confided in me. He was brilliant, driven, and visibly wrestling with the future. Leaning back in his chair, a stark contrast to the buzzing energy of the open-plan office around us, he described his company's new AI-powered loan recommendation engine. It was a marvel of machine learning, designed to accelerate approvals and reduce risk.

"Governance," he told me, a note of deep-seated frustration in his voice, "feels like a brake pedal. It's compliance paperwork, bureaucratic checklists, and endless meetings that kill momentum. We're building a rocket ship here, Omar. We have investors to answer to and a first-mover advantage to protect. The last thing I need is a team of auditors slowing us down."

His perspective was honest, and it's one I've heard countless times. But this conversation crystallizes the central theme of this chapter: that strategic leadership today is about orchestrating transformation at the critical intersection of artificial intelligence, governance, customer experience, and enterprise value. The rapid acceleration of AI conceals a deeper call: an invitation to embrace depth. As AI absorbs routine execution, leaders are summoned to cultivate a new kind of mastery. The ultimate differentiator is not the technology itself, but how we orchestrate it with human insight and ethical clarity. The essential question has shifted from *if* we will adopt AI to *how* we will govern and humanize it to create enduring value.

Governance: The Architecture of Enterprise Value in an Age of Uncertainty

For too long, we've treated governance as a cost center—a box to be checked. A cautionary tale shared with me by a leader at an HR tech firm perfectly illustrates the danger of that thinking. Her company, full of passionate people, rushed to deploy an AI tool to streamline hiring. Initially, the efficiency gains were celebrated. But a few months later, an audit revealed the algorithm, trained on historical data, was filtering out resumes with career gaps, disproportionately penalizing qualified women. The backlash was severe, eroding not just public trust but employee morale.

What this story reveals is that the risks of AI are not just about predictable biases. The deeper challenge is one of uncertainty. Unlike traditional software, which follows explicit rules, AI models are learning systems. Their behavior can drift. They can produce emergent outcomes that no one, not even their creators, anticipated. This introduces a persistent ambiguity into our operations. Governance, therefore, cannot be a static, one-time audit. It must be a dynamic, living process designed to navigate this inherent uncertainty.

This is where governance moves from a compliance mechanism to a philosophy of stewardship, the very foundation upon which enterprise value is built. Consider the fintech CEO I mentioned. He took a courageous step and reframed governance as a strategic enabler. His team established an ethics council, not just to approve models, but to wrestle with their ambiguity. He later told me about a moment of crisis that became a moment of clarity. Their model started flagging a certain type of small business loan as high-risk for reasons no one could immediately explain.

The old approach would have been to either trust the "black box" or discard the finding. Their new governance framework demanded a different response. The council initiated a "human-in-the-loop" review, where loan officers personally engaged with the flagged applicants. They discovered the AI was picking up on a subtle cash-flow pattern common to seasonal businesses—like a coastal seafood shack—that was a sign of normalcy, not risk. By investigating the uncertainty, they not only corrected the model but also built deeper relationships with a key customer segment.

A similar story unfolded in the even more highly regulated world of healthcare. A leader at a medical device company described the immense challenge of getting their new AI-powered diagnostic tool through the stringent approval process. The uncertainty was enormous: How do you prove the safety of a system that is constantly learning? Instead of treating the submission as a final, adversarial step, they orchestrated governance from the beginning. They built a transparent "explainability dashboard" specifically for regulators and meticulously documented their bias mitigation strategies. This act of radical transparency transformed a potentially multi-year, combative process into a collaborative review. The result was a faster approval, getting their life-saving technology to patients months, or even years, sooner. This is the intersection of governance and enterprise value in its purest form.

Customer Experience: The Human Intersection with Uncertainty

Nowhere is the need for orchestration more apparent than at the intersection of AI and customer experience (CX). The most profound CX strategies recognize that the goal is not total automation, but a harmonious blend of machine efficiency and human empathy, especially when facing the profound uncertainty of human emotion.

A leader at a major bank shared a powerful story that has stayed with me. For years, her operations team was focused on a single metric: average handle time. Yet, their customer satisfaction scores were stagnant. The breakthrough came when they discovered their highest satisfaction scores were not linked to speed, but to moments of deliberate slowness.

Their most impactful innovation was a "pause protocol" for account closures linked to a bereavement. The agent who led the pilot program later recounted the team's initial fear and uncertainty. "We're not grief counselors," she remembered one of her colleagues saying. "What if we say the wrong thing and make it worse?" The leadership team's role was to provide psychological safety and training, empowering them to navigate these unscripted, emotionally charged conversations. The agent described her first call under the new protocol. "I spoke to a man who had just lost his wife of 50 years. We didn't talk about the account for the first 10 minutes. We talked about her. He cried. I listened. At the end of the call, he thanked me for treating his wife like a person, not a file number." This single, empathy-driven change reduced complaints by 42% and fundamentally changed the culture of the support center. The team began to see themselves not as "ticket-closers" but as "relationship-builders," a source of immense professional pride. This is how customer loyalty—a critical component of long-term enterprise value—is truly forged.

The Conductor's Role: Leading Through Ambiguity

This brings us to the core of modern leadership. As AI handles more of the "what," our role shifts to defining the "how" and inspiring the "why." Strategic leaders are no longer commanders of efficiency; they are the conductors of transformation.

In an era of uncertainty, the leader's primary role is to be an anchor. The old model of leadership was about providing certainty—the five-year plan, the detailed roadmap. Today, when the technological landscape can shift in months, that model is obsolete. The new role of a leader is to provide clarity of purpose *within* the uncertainty. They don't pretend to have all the answers about how AI will evolve, but they are an unwavering source of truth about the company's values and mission. This leader is the one who says, "I cannot tell you exactly what our industry will look like in three years. But I can tell you that we will always prioritize our customers' trust and our employees' well-being. That is our North Star, and we will navigate by it, no matter how the technology changes." This approach builds the psychological safety that allows teams to innovate without fear.

A chief marketing officer I know embodied this when she realized her most important job was to be a "translator." Her data science team would present their latest customer churn model, proudly announcing its "AUC score of 0.92." Meanwhile, her brand managers would stare back blankly. The two groups were speaking completely different languages, creating a culture of confusion and mild distrust. Her solution was to ban technical jargon from strategy meetings. She asked the data team to present their findings as "customer stories." The "high-risk segment" became "our loyal families who feel their needs are changing." The meetings transformed. By creating a shared language, she navigated the interpersonal uncertainty and orchestrated a harmony between two previously disconnected parts of her organization, reducing churn by 15%.

Finally, leaders must lead with empathy and ethical clarity. A leader at a traditional manufacturing company faced this head-on when introducing robotics. Instead of a technical presentation, she held a town hall. "These robots are not here to replace you," she told her team. "They are here to liberate you from the dangerous, repetitive work, so you can become the strategists and innovators who run the factory of the future." In that moment, she was orchestrating transformation, wielding the narrative baton to turn the uncertainty of the future into a shared sense of purpose.

Final Thoughts: The Invitation to Mastery

I sometimes think about that fintech CEO who saw governance as a brake pedal. His journey reminds me that true leadership isn't about having all the answers; it's about having the humility to navigate the unknown. The future of work is not a sterile landscape of algorithms. It is a deeply human terrain, one that calls for a new kind of leadership that is comfortable with ambiguity.

We stand at a profound paradox. Artificial intelligence offers a seductive illusion of control, a world of data-driven certainty and flawless optimization. Yet, in practice, its implementation introduces a new and persistent form of operational, ethical, and strategic uncertainty. The role of the modern leader is to hold this tension: to leverage the power of the machine while honoring the wisdom of human intuition, to pursue innovation with courage while grounding it in an unwavering ethical framework.

The conductor does not play every instrument; their genius is in drawing out the best from each section to create a cohesive whole. The architect does not lay every brick; their value is in designing a structure where people can thrive. Similarly, our role as leaders is not to become expert coders or data scientists. It is to become expert humanists. It is to orchestrate the immense potential of our technology in a way that amplifies the best of our humanity.

Artificial intelligence is not our competitor. It is a catalyst for our own rediscovery. By absorbing the routine, it clears space for us to cultivate mastery and purpose. It compels us to refine what makes us unique: our nuanced ethical judgment, our profound emotional intelligence, and our ability to lead complex change in the face of persistent uncertainty.

Herein lies AI's hidden invitation. It is an invitation not to a race against the machine, but to embrace the distinctive qualities that make us human. It is an invitation to elevate our collective potential by placing the human element at the very heart of all innovation. This is the era of depth. Our role as leaders is to orchestrate the intersection of AI, governance, and customer experience in a way that creates authentic, sustainable enterprise value. The legacy we leave will not be measured by the technology we deployed, but by the human capacity we unleashed. Our answers to this call will define not only our competitiveness but the very character of leadership for decades to come.

References

Boston Consulting Group. *Bridging the Gap Between Data Science and Business Strategy*. Boston Consulting Group, 2023.

Boston Consulting Group. *AI and the Age of Mastery*. BCG Insights, 2025.

Credo AI. *Mastercard and Credo AI: Scaling Generative AI Governance in the Financial Sector*. n.d. Credo AI.

Deloitte. *AI Governance: The New Frontier of Risk and Value*. Deloitte Review, 2024.

Gartner. *AI Orchestration as a Leadership Competency*. Gartner Research, 2025.

Google Cloud. *Real-World Gen AI Use Cases from the World's Leading Organizations*. n.d. Google Cloud.

GovNet. *How Governments Are Using AI: 8 Real-World Case Studies*. n.d. GovNet.

IBM. *What Is AI Governance?* n.d. IBM.

McKinsey & Company. *The State of AI 2024: Scaling with Governance*. McKinsey Global Institute, 2024.

McKinsey & Company. *Personalization at Scale: AI in Customer Experience*. McKinsey Global Institute, 2025.

PartnerHero. *5 Real-World Examples of AI in Customer Experience*. n.d. PartnerHero.

PwC. *Experience Is Everything: Here's How to Get It Right*. PwC, 2024.

Relyance.ai. *AI Governance Examples—Successes, Failures, and Lessons Learned*. n.d. Relyance.ai.

University of Toronto Libraries. *A Case Study for AI Governance*. n.d. University of Toronto Libraries.

Zaman, Kashif. *Agentic Leadership: Timeless Truths, New Intelligence*. 2025.

Zendesk. *13 Ways AI Will Improve the Customer Experience in 2025*. n.d. Zendesk.

About the Author

Mohamed Omer is an AI leader and strategic advisor who helps organizations navigate the complex intersection of technology, governance, and enterprise transformation. After beginning his career focused on operational strategy, he discovered that the most significant challenges in technology were not technical but human—a realization that prompted him to focus on developing human-centric AI strategies. He believes that true, sustainable value is created when innovation is anchored in trust and purpose. Mohamed is a passionate advocate for responsible innovation and believes that the true power of AI lies not in replacing human effort, but in augmenting human potential to solve our most meaningful challenges.

Email: maomer@gmail.com

LinkedIn: https://www.linkedin.com/in/mohamed-omer-mba/

CHAPTER 25

DIGITAL DARWINISM: THE NEW RULES OF SURVIVAL IN AN AI WORLD

By Pramod M. Patke
Product Manager, System Architect
Gothenburg, Sweden

> *It is not the strongest of the species that survives, nor the most intelligent, but the one most adaptable to change.*
>
> —Leon C. Megginson (inspired by Darwin)

Picture this: It's early morning, the city's still draped in blue-grey twilight. On the outskirts, there are steel bones of an old factory, a faded sign. It's the kind of place generations called their livelihood, and it creaks awake. Inside, workers in heavy boots clock in, their routines unchanged for decades. The line starts rolling, sparks fly, and there's comfort in the familiarity, but also a restlessness, an unspoken worry about how long this world will last.

Now, shift your gaze just a few miles down the road. There, in the shell of a warehouse that once made engine blocks, a new kind of factory pulses with quiet, purposeful energy. Floor-to-ceiling windows let in the first hints of sunlight. Agile robots zip along marked pathways, carrying out a silent choreography. Human technicians, some in hoodies, some in lab coats, stand by digital dashboards, their focus not on repeating the same motion a thousand times, but on solving problems, fine-tuning algorithms, and collaborating with their silicon coworkers. In this place, decisions are informed by a stream of live data about parts, performance, even the emotional pulse of the team. The day's first challenge? Not how to keep up, but how to stay ahead.

Between these two worlds, the familiar comfort of yesterday and the pulse of tomorrow runs a fault line. It isn't about machines replacing people; it's about what happens when change itself becomes the only constant.

This is digital Darwinism in real time. Across industries and continents, in boardrooms and on shop floors, everyone's asking the same thing: Who adapts and who gets left behind? Yet, unlike evolution in the wild, we aren't just passengers on this journey. We get to choose how we respond and what kind of future we create.

The real question isn't whether AI will change the world (it already has). The question is, as these tides rise: Will we drift, or will we steer? In this new survival game, the winners won't simply be the ones who deploy the most technology, but those who blend adaptability, vision, and human values into everything they do. Because when the dust settles and the world looks back at this moment, the most remarkable thing won't be what our machines could do. It will be what we chose to do with them.

Winners and Losers: The New Rules of Adaptation

The illiterate of the 21st century will not be those who cannot read and write but those who cannot learn, unlearn and relearn.

—Alvin Toffler

Adaptation isn't an event; it's a perpetual state of mind. The most striking feature of this new landscape? Winners and losers are no longer determined by size, age, or even wealth, but by the speed and wisdom of their response to change.

Consider the automotive industry, once a monument to routine, now a test lab for survival. Some automakers bet big on electric and autonomous vehicles, transforming their factories, retraining their people, and even reimagining themselves as "mobility companies." Others, slow to move, clung to their old combustion playbook only to watch market share erode, talent slip away, and headlines turn sour. The message? In the AI economy, tradition is not a moat. Reluctance to evolve is an anchor. But the same story unfolds everywhere:

- In finance, nimble fintech startups use AI to spot fraud and serve customers 24/7, while banks with sprawling branch networks and legacy IT struggle to catch up.

- In healthcare, clinics leveraging machine learning for diagnosis and patient management deliver faster, more accurate care while those stuck in paper and procedure see waiting rooms swell and patients look elsewhere.

- Retailers that let algorithms guide supply chains and personalize customer experiences are thriving; those slow to digitize see empty aisles and shuttered storefronts.

- In agriculture, farmers using AI-powered sensors to monitor soil and predict harvests outpace neighbors, relying on "how it's always been done."

It's the same evolutionary game, replayed on every continent, across every industry: Adaptation is king. Yet, it's not just about acquiring shiny tech. The real winners are those who cultivate a mindset of continuous learning and ethical action, who see every disruption as a chance to reimagine what's possible.

Look closer, and you'll see something else: This new adaptation isn't just technical. It's deeply human. The people who thrive are those willing to challenge old assumptions, to "learn, unlearn, and relearn," as Toffler put it. Organizations win when they reward curiosity, encourage experimentation, and welcome dissent, not just efficiency and conformity.

On a global stage, the gap widens. Countries and companies that invest in upskilling their people, in open data, and in inclusive AI policy are building ecosystems that attract talent and capital. Those who merely defend the status quo find themselves left behind, not just by their competitors but by the very pace of history.

It's not just business that's evolving. Governments around the world are deploying AI to improve everything from traffic management to digital public services. Estonia, for instance, has automated over 60 government processes using AI. In education, countries like Singapore and Finland have launched national AI literacy programs for students and teachers, while major US school districts pilot adaptive learning platforms that personalize instruction for millions of learners.

Quantitative Snapshot: The Scale of Change

According to UNESCO's 2024 Global Education Monitoring Report, public investment in AI-driven education technologies topped $12 billion last year. A 2024 IDC study forecasts that global government spending on AI will exceed $40 billion a year by 2026. The fastest growth is in e-governance, urban mobility, and healthcare.

The World Economic Forum reports that nearly 44% of workers worldwide will need reskilling or upskilling by 2028, driven by AI automation. In rapidly changing fields like automotive and finance, that figure climbs above 60%. Meanwhile, a World Economic Forum 2024 survey shows that nearly 44% of workers globally will need reskilling or upskilling by 2028 due to AI-driven automation, a figure that rises to over 60% in fast-evolving industries like automotive and finance.

In digital Darwinism, everyone is a startup. No one is immune. The good news? The same forces that threaten to disrupt also offer a ticket to renewal. Adaptation isn't a burden. It's a passport to a future that's ours to invent.

Philosophy in the Machine: Purpose, Consciousness, and Human Uniqueness

Beneath the whirlwind of AI advances, machine victories, market disruptions, and dazzling inventions runs a more subtle, personal reckoning: If machines can learn, plan, and even create, what's left that is truly ours? For all its circuitry and code, the rise of AI drags ancient questions back into the limelight: What makes us human? What is the role of meaning, intuition, and ethics in an age of thinking machines?

Techno-optimism asks us to see possibility in this uncertainty, to view AI not as an eraser of purpose, but as a catalyst for rediscovering it. History's pattern is reassuring; with every leap in technology, humans have found ways to shift from repetitive survival tasks to creative, connective, and meaningful pursuits.

But something deeper is happening in today's AI revolution. Geoffrey Hinton, widely known as the "Godfather of Deep Learning," has argued that AI is not following the logic of traditional reasoning. In his Nobel Prize acceptance, Hinton reflected that the astonishing power of today's neural networks stems from their ability to intuit rather than strictly reason. AI learns from vast rivers of data, finding patterns too subtle or complex for step-by-step logic, much like a human's gut feeling or a chess master's instinctive move.

This brings both opportunity and danger. On the one hand, intuitive AI enables breakthroughs: spotting rare disease in a scan, predicting supply chain shocks, preventing a car crash in milliseconds. On the other hand, Hinton warns, such intuition-driven systems can easily reinforce our own blind spots. And the risks, he adds, aren't just short term. Without careful stewardship and ongoing research, these systems could amplify bias, polarize society, or even drift out of human control.

This reality reframes what it means to win in digital Darwinism. The advantage no longer lies simply in deploying more AI, but in understanding and guiding it. We must ask: What kind of intelligence do we want to create, and what human purposes should it serve? If meaning the "why" behind the "how" remains our domain, then our job is to ensure that technology deepens, not diminishes, what matters most. The greatest promise lies in the fusion of humans and machines. The new edge is collaboration, not replacement.

As organizations race to automate, the true winners will be those who keep the human in the loop: upskilling workers, questioning outcomes, and ensuring that technology remains a servant of authentic human need. Because in this era of digital Darwinism, our real advantage will not just be adaptability, but the wisdom to remember what matters and the courage to demand more research and reflection before rushing ahead.

AI Without Data: Thriving in Data-Starved Environments

In the real world, most organizations are working not in abundance, but in scarcity: limited data, messy records, edge cases the dataset barely covers. Here, success belongs not to those with the biggest servers, but to those who innovate when resources are thin.

Take the automotive sector. Training a self-driving car to handle every possible road hazard, weather anomaly, or pedestrian movement would require more driving data than any company could realistically gather. The clever trick? Automakers have learned to combine real-world data with simulated scenarios of digital twins of city blocks and rural roads, where rare dangers can be "experienced" thousands of times before a single tire hits the ground. This approach lets cars learn not just from what *has* happened, but from what *could* happen.

A similar pattern shows up in healthcare. Rare diseases don't come with millions of patient records, but new AI tools use what's called transfer learning: They borrow what's learned from common conditions and apply it to the rare, adapting old knowledge to new problems. In agriculture, a single season crop data can be bootstrapped with weather models and expert intuition to create early-warning systems for farmers, helping them anticipate drought or pests, even when historical data is thin.

There's a lesson here: The future isn't about who has the most data, but who makes the smartest use of whatever data they have. In fact, scarcity can be an advantage. With less data, teams are forced to focus on the right questions, what really matters, what signals are truly useful, and what risks can't be ignored. They combine different types of knowledge: human intuition, simulation, domain theory, and sometimes even synthetic data generated from scratch.

Ethics as Strategy: Evolving Consciously

In the first wave of digital disruption, speed was everything. But as AI weaves itself deeper into decisions that affect lives, economies, and even democracies, a new kind of edge is emerging: the ability to navigate complexity with a moral compass.

Ethics isn't just a checkbox or a public relations move; it's increasingly a driver of sustainable competitive advantage. Organizations that treat ethical considerations as central to their AI strategy build trust with customers, employees, and society at large. Those that ignore them may win for a while, but risk public backlash, regulatory headaches, or even irrelevance if people lose confidence in their brand.

Ethics in Action: Lessons Learned

When a major US retailer discovered bias in its AI-powered hiring tool, it paused deployment, published findings, and invited external audits—ultimately regaining public trust and improving candidate diversity. In contrast, a global ride-sharing company ignored early warnings about its facial recognition AI, leading to lawsuits and a ban in several cities.

This shift is global. The European Union is rolling out some of the world's strictest AI regulations, focusing on transparency, accountability, and human oversight. Companies in Asia, North America, and Africa are developing their own ethical frameworks, not out of obligation, but because they see that fairness and explainability make technology more usable and more accepted.

There's a deeper point here, too: Ethical thinking pushes companies to ask, "What is AI really for?" Instead of just optimizing profit or efficiency, forward-looking organizations consider broader impacts on how AI shapes society, who benefits, and who might be left out. The winners in this new game will be those who prove, day after day, that AI can be powerful *and* principled.

Automotive as Microcosm: The Fast Lane of Evolution

If you want to see digital Darwinism up close, look under the hood of the modern auto industry. This sector, once a symbol of predictability and routine, has become a proving ground for every lesson of the AI age.

For decades, carmakers measured success by how many vehicles they could assemble and ship. Now, survival hinges on something very different: the speed at which companies can reimagine what a car and a car company should be. Some brands saw the writing on the wall early. Tesla approached vehicles as "computers on wheels," making software updates central to steel and rubber. Their willingness to treat the car as a digital platform, not just a product, puts pressure on giants like GM, Volkswagen, and Toyota to adapt or risk obsolescence.

Inside the factory, the shift is just as dramatic. Smart robots and machine vision systems don't just speed up production; they turn the assembly line into a living, learning organism. Algorithms predict when a part might fail before it halts the line. Workers find themselves stepping into new roles as process designers, AI troubleshooters, and data interpreters, jobs that didn't exist a decade ago.

On the road, the stakes are even higher. Cars now "see" their environment using networks of cameras, radar, and AI-driven perception systems. The race to develop safe, trustworthy autonomous vehicles isn't just a technical challenge; it's a question of public trust and global leadership. Who will write the code that decides how a car handles the unexpected, a darting pedestrian, a blocked intersection, a sudden storm? The answer will shape not just brands and market share, but the safety and livability of whole cities.

Globally, the transformation looks different in every region. In China, electric vehicle adoption has surged, driven by urban innovation and government support. In Sweden, smart mobility projects link self-driving shuttles with public transit, while in Africa, mobile-based logistics platforms powered by AI help move goods to the last mile. In every context, those who thrive are not just chasing the latest tech, but are tuning in to local needs, regulatory realities, and the human experience of mobility.

The auto industry's evolution also highlights a crucial theme: Adaptation isn't just for companies; it's for communities and workers, too. Factories are being retooled, but so are careers and skills. Those who invest in retraining, who partner with schools and startups, are building ecosystems that can absorb shocks and seize new opportunities. In the end, the automotive story is a mirror for every sector touched by AI. It shows that the future belongs to those who are willing to reimagine, relearn, and rebuild, not just products, but the very roles and relationships that keep an industry alive.

The Human Playbook for Thriving in AI's Age

In the AI era, lasting success is within reach, not just for the biggest or the fastest, but for those who act with intent on individuals, leaders, and organizations alike. Recent breakthroughs and global strategies make the path clearer than ever.

Lifelong Learning

Thriving today means treating learning as a continuous journey. Recent years have seen major organizations invest billions to equip people with AI skills. The message is simple: Those who make upskilling and curiosity part of daily habits are best positioned to seize new opportunities and adapt to change.

Human and AI Collaboration

Far from replacing people, AI is increasingly a teammate. By 2025, "AI copilots" are embedded in everything from office software to industrial design. Studies show that workers who actively embrace AI collaboration are not only more productive but also more satisfied, turning their teams into what analysts now call "superworkers."

Ethical Courage

With new power comes new responsibility. Recent regulations like the European Union's AI Act and the US Executive Order on AI are making it clear: Organizations must treat ethics as a core competency, not an afterthought. Leaders now empower teams to question and pause risky AI deployments, conduct bias audits, and ensure "human-in-the-loop" oversight because trust and reputation are now just as important as speed.

Global Mindset

The AI revolution is borderless. Today's innovators build cross-cultural teams and adapt solutions for a patchwork of local rules, from Europe's transparency standards to China's content labeling laws. Companies like Toyota are uniting engineers from Japan, Europe, and the Americas in their global AI accelerators, proving that international collaboration is a multiplier for fresh ideas and resilient solutions.

Purpose-Driven Adaptation

The most inspiring organizations and individuals are tying their AI ambitions to a clear sense of purpose. Microsoft's AI initiatives are focused on accessibility, education, and "AI for good." Toyota's strategy links every AI investment to freedom of mobility and quality of life.

In practice, this means not just building powerful tools, but ensuring those tools are used for meaningful, human-centered progress.

Above all, those who thrive in this era are guided by clarity of purpose. They use change as a lever not only to do more, but to do better for themselves, their companies, and their communities.

Facing the Messiness of Change: Inertia, Missteps, and the Human Factor

It's tempting to present digital Darwinism as a story of rapid adaptation and clear winners. In reality, change is rarely tidy. Both in boardrooms and on factory floors, organizations hit old habits, resource shortages, and cultural resistance that slow or even derail progress.

Take reskilling. Even with robust programs, many companies find that not everyone can or wants to make the leap to a new digital role. For those left behind, the promise of a high-tech future can quickly turn to frustration and loss.

Organizational inertia is another major obstacle. Legacy systems, entrenched routines, and fear of the unknown cause even market leaders to delay tough decisions or underinvest in new skills.

Culture is always a wild card. In some organizations, risk aversion or rigid hierarchies can sap the energy from innovation before it has a chance to take root. Cross-functional AI teams can stumble if trust and communication aren't actively nurtured. Ethical lapses become more likely when diversity and dissent are missing.

These pitfalls don't make adaptation impossible. But they do make it slow, messy, and deeply human. Success in digital Darwinism depends as much on humility and persistence as it does on vision and speed.

Reflection: The Evolution We Choose

The future depends on what you do today.
—Mahatma Gandhi

For most of history, evolution was a blind process of life adapting by accident, trial, and error. But in this new digital age, evolution is something else entirely: a conscious act. We are no longer just subjects of change; we are its architects, its authors, its co-creators.

Every algorithm, every new workflow, every bold experiment is a reflection of our values. In a world that rewards adaptability, the true winners are not those who simply keep up, but those who choose to move with purpose. The most lasting progress will come from leaders and communities who evolve with clarity about what matters most, those who pair curiosity with conscience, innovation with empathy.

So, yes, digital Darwinism can feel unforgiving, but it is also an invitation. To reassure, to inspire. The real opportunity is not just to survive the whirlwind of change, but to shape it deliberately, wisely, humanely. The future is not being written for us; it is being written *by* us. As Gandhi reminds us, "The future depends on what you do today." Let's make today's choices count. As we step forward, what future will you help invent?

About the Author

Pramod Patke is a visionary technology leader, strategy product manager, and startup founder with over 15 years of experience driving AI transformation across industries worldwide. He has partnered with leading organizations to advance intelligent systems in fields ranging from automotive and manufacturing to public sector innovation and digital education.

As an expert in autonomous systems architecture, data-driven decision-making, and machine learning, Pramod bridges the gap between advanced engineering and business strategy. His entrepreneurial mindset has shaped the launch of new ventures and scalable AI solutions for industry, while his strategic leadership fosters ethical technology adoption and lifelong learning in the workforce.

Pramod is also co-author of the book *AI Revolution*, where he explores the transformative impact of artificial intelligence on organizations, society, and the future of human-machine collaboration. Passionate about a future where AI benefits all, he inspires professionals and enthusiasts to navigate the AI-driven era with curiosity, courage, and conscience.

Email: pramodpatke@gmail.com

LinkedIn: www.linkedin.com/in/pramodpatke

CHAPTER 26

INDUSTRIAL AI UNLOCKS HIDDEN POTENTIAL

By Robert Pluska
Industrial AI Strategist, Advocate, and Founder of Advanced Solutions at JSP
Dusseldorf, Germany

> *Innovation is never about technology. It's about solving a customer's problem.*
>
> —*Clay Christensen*

AI and Manufacturing

The most significant missed opportunity in manufacturing today isn't a lack of data. It's failing to utilise the knowledge you already possess. AI isn't limited to high-tech labs or futuristic factories. It's already helping traditional manufacturers improve quality, reduce waste, become more sustainable, make smarter decisions, and act faster—right on the shop floor.

Still, many people view AI from the wrong perspective. Too often, industrial AI is mistaken for flashy marketing tools, gimmicky chatbots, or fully autonomous machines. That perception creates a gap. It causes leaders to think AI is either irrelevant or unrealistic.

Industrial AI is not a far-exaggerated concept. It's a practical tool that doesn't aim to replace people, but to enhance their capabilities. It's about capturing the knowledge of experienced workers, facilitating quicker learning, and identifying problems before they escalate. It's not just about data, it's about understanding. It doesn't require perfect data or flawless systems. It needs context, clarity, and connection. This practicality should reassure you of its potential in your manufacturing operations.

The most significant opportunity in industrial AI isn't creating something new. It's unlocking what companies already know. In most factories, valuable knowledge is hidden in spreadsheets, siloed systems, paper logs, or the minds of experienced workers nearing retirement.

I've seen teams spend days searching for reports or repeating costly experiments to rediscover known facts. I realised that industrial AI's core mission is to organise and activate internal knowledge so that it's accessible, helpful, and available for informed decisions. It prevents endless rediscoveries through folder searches or trial and error. AI provides synthesis, accessibility, and action, even as teams change.

Just as OpenAI captures general knowledge, industrial AI should organise what your factory knows behind your firewall. Present it to reduce waste, improve safety, and support decisions. Transformation begins with organising existing data, not with algorithms. This chapter explains how we introduced AI into outdated manufacturing processes to make them practical, personal, and valuable. I believe AI's most significant industry value isn't replacing workers but supporting them.

The Dashboard Trap

Industry 4.0, the fourth industrial revolution, was introduced in Germany in 2011. It represents a significant shift in manufacturing, characterised by the integration of digital technologies and the internet of things (IoT) into the production process. Yet, over a decade later, many manufacturers remain stuck at the dashboard stage. ERP and MES systems are installed, control panels are mounted on walls, and KPIs are tracked. But real progress often ends there. Many of these tools are like multitools—capable of doing everything but rarely used effectively. Usually, what's needed isn't another feature but a precise tool tailored to a real problem. This is where AI offers something different: not more data, but better decisions.

I've seen dashboards in every plant, including our lab, but most are ignored. If everything looks fine, no one checks them. If something goes wrong, people scramble into crisis mode. The dashboard becomes irrelevant; it didn't prevent the issue and doesn't guide recovery. It just sits there.

AI, in contrast, is not a passive observer. It's an active problem-solver. It monitors patterns, detects early warnings, and highlights what is essential before things stray from their specifications. This marks a significant shift: from ordinary visibility to taking action. This shift should empower you, as it means you're not just seeing what's happening, but you're in control and can take proactive steps to prevent issues.

Interoperability as a Foundation for AI

One of the most valuable long-term projects I've worked on had nothing to do with directly deploying AI. It was all about enabling future AI by creating a shared data foundation. The initiative started with a simple yet frustrating challenge: Each machine manufacturer in our industry had its data format, naming conventions, and interface protocols. As a result, collecting usable production data from different machines—even on the same factory floor—was discouraging.

We decided to change that. Alongside three leading machine makers in the particle foam industry, we started developing an OPC UA Companion Specification for our industry. Initially, the conversations were slow and cautious. Each supplier had its legacy systems, naming conventions for variables, and, in some cases, resistance to standardisation.

It took two full years. Eventually, through dozens of workshops, mutual alignment, and continuous iteration, we reached a shared protocol—a single "language" that machines from different suppliers could use to communicate in a unified and interoperable way. This wasn't just a one-time hack for integration; it was a comprehensive solution that provided a lasting benefit. It became a public standard, openly available for the entire industry.

The immediate result was technical: Connecting machines became easier, and data collection grew simpler. But the broader impact was strategic. This interoperability unlocked use cases we couldn't access before. Suddenly, we could monitor utilities, cycle times, and maintenance events across the entire machine fleet, including both old and new machines. Such connectivity laid the groundwork for AI agents that require consistent, timestamped data streams.

Case Study: Phase One—From Fragmentation to Aggregation

One of my most fulfilling projects started with a question: "Can we connect all our data to see what is happening?" It came from a German manufacturing CTO with technical expertise and a business vision. After observing our lab work, he saw the potential to integrate it into his new OPC UA-enabled production plant.

Before we started, the factory collected data across various databases—machine logs, energy meters, and utilities—creating silos. Operators identified inefficiencies, including compressed air losses and energy spikes. Diagnosing issues was slow due to the manual nature of experiments and data comparisons. We partnered with the customer to create a unified data architecture utilising OPC UA, consolidating high-frequency raw data into a single system. This enabled the plant to monitor all processes in real time, replacing the need for spreadsheets and isolated reports.

The team finally pinpointed when and where losses happened. The first breakthrough occurred after the data sync, when the CTO noticed compressed air leaks on the dashboard, revealing waste. "We've known we lost money here," he said, "but couldn't prove when or where." This led to estimated six-figure annual savings. Their CO_2 calculator tracked every kilowatt-hour saved, linking process gains to emissions reductions in real time. It showed how digitalisation boosts business and supports sustainable development goals.

Beyond visibility, the unified data architecture enhanced security and resilience by implementing encryption, data compression, certification, and authentication, integrating cybersecurity from the outset. Backup and recovery functions protect data in the event of network disruptions. The project's success wasn't just about technology, but also about culture, with no resistance, just curiosity about the results. Starting with a focused problem and proven tools, we quickly gained momentum.

Case Study: Phase Two—From Data to AI

After the integration phase, we proceeded to the second stage: shifting from data visualisation to providing actionable decision support. We assisted the customer's team in understanding the concept of industrial AI agents—software that not only displays information but also learns from it and acts on it. Since we had already developed AI agents internally, we were able to swiftly incorporate this capability into the project, just as we did with the data architecture.

The idea of an AI agent that could cross-reference numeric process data with contextual knowledge—such as operator input, maintenance logs, and historical trends—resonated strongly. The team immediately understood how it could deliver clear recommendations or automate repetitive decisions that previously required a significant amount of time and attention. It was a moment when everyone realised that this project could become a benchmark in the industry, a clear example of how strong foundations enable the scaling of successful solutions even into less sophisticated production environments.

If I had to advise anyone on a similar project, I would say: Assign a senior leader to inspire and clarify priorities. Be patient—ROI takes time. Select partners who add value early. Connect existing assets before creating new ones. This project confirmed that interoperability is essential for meaningful AI change; without it, models struggle in the dark.

People and Trust

The reason we made real progress with AI in manufacturing by 2025 was that we had spent years thinking carefully, not just about how to build the technology itself, but also about how to lay the foundations and present its potential in a way that people could trust. We worked on data architecture, experimented with AI agents, and closely watched how teams responded. Great tools alone aren't enough. If people don't trust them, nothing changes. That's why culture matters just as much as technology. We learned this very early.

The Day AI Polarised the Room

When GPT arrived in 2022, I knew we needed to explore its potential, not just technologically, but culturally. We rolled out a simple internal demo without hype. We aimed to demonstrate our understanding of the technology and its application with a focused knowledge base. Reactions were fascinating: some curiosity, "Can it generate instructions from machine data?"; others a bit uneasy, "Will it replace us?" And a quiet group, interested but cautious. That demo didn't change minds overnight, but it made AI real. It's no longer just a trend; it has raised questions and some tension, which matters.

Since then, we've shifted our AI approach to focus on usefulness rather than promises. Instead of claiming, "This will change everything," we say, "It might help you tomorrow," fostering curiosity rather than resistance. AI adoption depends on timing and trust; if people feel excluded, even the most innovative tech fails. If they see it as a tool that adapts to them, they'll embrace it. A simple demo revealed more about our readiness than any framework, reflecting our culture and shaping our AI rollout through dialogue, not just code. The GPT experience was just the start. We've since applied this mindset to practical industrial AI tools, starting small, learning fast, and solving real problems.

Trust Starts with Listening

These lessons about perception shaped how we approached every new project. When we started working with a European automotive supplier, we applied the same mindset: Keep it simple, make it worthwhile, and involve people early.

The factory was attempting to enhance the effectiveness of its equipment and stabilise the production of lightweight particle foam parts. These processes were sensitive to temperature and steam fluctuations, so any downtime disrupted quality and profitability.

What made this case stand out wasn't the complexity of the challenge. It was how disconnected everyone felt from the data. Machine logs, shift reports, utility records, and quality checks existed but were scattered across folders, disconnected systems, and handwritten notes. Just gathering them took an hour each day, and even then, insights often arrived too late to be acted upon.

Our first step wasn't to build anything. It was to bring everyone together. We sat down with operators, engineers, and managers—not to pitch a solution, but to ask questions and listen to their perspectives. The team shared a list of frustrations and recurring questions: "Why does scrap spike on night shifts?" "Can we spot problems before they escalate?"

Only after that conversation did we begin designing a monitor application to consolidate their information in one place. Because the questions came from them, the tool felt familiar. The goal wasn't to impress them with technology; it was to show them that their experience mattered.

A Reliable Assistant Deserves a Name

Even after enhancing data visibility, most logs were still stored and forgotten because reviewers lacked the time to review hundreds of entries weekly. Operators were stretched thin, and more dashboards wouldn't help. That insight became the impulse behind our first practical AI assistant. We called it "Jose." The idea was simple: Instead of creating another reporting layer, we wanted something that could sit inside existing workflows and quietly help prioritise what mattered most.

Jose wasn't designed to be impressive. It didn't predict the future or replace human expertise. It simply read log entries continuously, calculated which issues cost the most, and displayed a single clear recommendation: "Focus on this one."

Although Jose was developed after the automotive supplier project, it was based on the principle that trust grows when tools respect existing workflows. Instead of drastic changes, we focused on small, natural steps. This mindset guided all future AI deployments.

For me, this was proof that real progress doesn't require complex systems. It requires listening first, then building something that fits. These experiences shaped how we think about scaling. When leaders ask how to start, I share the same approach: Start small, build trust, and focus on problems that matter.

Act Before You're Forced To

If you ever wonder, "Should we wait for perfection?" I suggest, "Begin by addressing what's already faulty. That's the way to develop readiness." Digital transformation isn't about finding perfect tools, but using the right ones. Begin by addressing a single problem or process. Don't try to do everything at once. Work with your operations team, not just IT, to identify issues and improve adoption. Set realistic expectations. AI can't fix everything, but it can boost speed and accuracy.

One example of factors influencing the success of digital initiatives involves two pilot projects launched simultaneously to reduce scrap and improve process stability on block production lines; however, their outcomes differed. The first team operated 30-year-old machines without modern sensors but was proactive, establishing a simple data logger where operators recorded downtime and defects. Although manual, the process was well organised, and the data uncovered links between process conditions and failures, revealing patterns no operator could see alone.

The second pilot had advanced tech, IoT, and dashboards, but expectations were misaligned. Data was collected, yet there was no consensus on interpretation or action, and operators felt disconnected; thus, the system lacked traction. This shows that success isn't about the newest machines, but rather a clear purpose, ownership, and willingness to start imperfectly. The first team saw their pilot as a stepping stone, while the second saw it as a showcase. Only one delivered results. Treat pilots as stepping stones, not just checks on a list. Design them to scale if they work. Most factories already have the data; connecting it in new ways yields the best results. Invest in your people, too, not just new tech. If your team doesn't have the time or confidence, even the best solutions can fail.

Share successes to foster innovation. Digital transformation isn't risky if it's relevant and supported. Create clarity yourself, or the environment will force change upon you. The gap between early adopters and hesitant firms is widening.

The Future Incremental Innovation

The future of industrial AI will evolve gradually through interconnected steps—machines that communicate, systems that learn, and teams that adapt. In the particle foam industry, traditional machines have developed into "fusion products", recording performance, monitoring energy use, or sharing data and connecting to fusion services like predictive maintenance and real-time advanced analytics, reducing downtimes.

The next focus in our development is fusion systems, which not only allow connected machines to collaborate but also gather data to produce insights and interpret them to support well-informed decisions. Evolution includes interfaces beyond screens, such as intuitive prompts, with AI aiding operators to learn more quickly and manage with greater confidence, thereby bridging the workforce gap.

Edge computing will enable AI to operate offline near factory equipment, processing data where speed and reliability are most critical. This will help manufacturers scale industrial AI with greater confidence. Production teams will access platforms to develop AI agents tailored to their workflows and customers within digital supply chains. This transformation is underway in pilot projects and early rollouts. Progress takes time, but starting small will accelerate success.

Conclusion

Manufacturing needs better decisions, not dashboards, by connecting people, systems, and data thoughtfully. AI helps reduce scrap, stabilise processes, and share knowledge, but success depends on how it's applied—purposefully, contextually, with people in mind.

Future thriving companies will leverage existing tools effectively, rather than relying solely on the latest technology. True digital transformation isn't just about technology; it's what manufacturing can achieve with previously unimaginable technology.

Start small: Solve problems, support teams, and generate insights. Transformation begins with relevance, not scale. Solutions matter little if people don't feel a sense of ownership. Start where you are, use what you have, and trust that clarity outweighs any roadmap.

About the Author

Robert Pluska was born in Warsaw, Poland, and is a manufacturing strategist with over 20 years of hands-on leadership experience in digitalising industrial processes. He specialises in helping small and medium-sized manufacturers unlock value through practical, data-driven solutions that align with the sustainable development goals. Robert led the creation of Dynamic Data Exchange (DDE Technology), a digital platform for particle foam applications, and transformed it into a standalone business unit—Advanced Solutions—within JSP. His approach combines strategic thinking with operational pragmatism, enabling companies to modernise without disruption. He is a strong advocate of agile transformation as a pathway to digital maturity, focusing on solutions that support people, not replace them. Whether the goal is to reduce waste, improve energy efficiency, or secure knowledge before it's lost, Robert brings clarity, structure, and momentum to the adoption of industrial AI.

Email: rpluska@gmail.com

Website: www.visionaryengine.de

LinkedIn: www.linkedin.com/in/robertpluska

CHAPTER 27

A ROAD TO AGENCY: FROM MULTIMODALITY TO SUPERINTELLIGENCE

By Niyati Prajapati
AI Lead and Tech Executive
Mountain View, California

The future is already here; it's just not evenly distributed.
—William Gibson

The Edge to AI Revolution

Overlooking the map of human civilization, the technological arc started from steam engines, punch-card computers, and blinking lights, which defined the mechanical and digital age. Over the past three and a half decades of the internet and intelligent age of smart sensors, vision robots are now eventually approaching the crossroads of agentic AI and superintelligence.

Each new advancement of this era has reshaped and redefined social, economic, and industrial spaces to push humanity at new innovation peaks. The upcoming shift from intelligence to superintelligence represents the next leap, one that demands a new level of stewardship.

As machines amplified physical labor, urban immigration, industrial capitalism, and manufacturing, there was a plateau at energy and cognitive constraints. When the complexity of logic mechanized, the microchip was named as an economic multiplier, but it was still lacking large scale connectivity and hardware speed. When globalization accelerated through the internet, social media governed digitally enabled business models. The AI understood near infinite information gathered from images, text, and speech to generate brand-new ideas and became a specialized force in supply chain optimization, robots, and autonomous vehicles by inheriting new value from training data and foundational models.

Since then emerged the human-supervised agents, multimodality and self-improving systems easing cognitive human load by condensing the time needed for tasks from decades to minutes. In addition to abundant value creation, AI governance and ethical steering, human design experience, relevance, and alignment mattered more than before to drive distribution and power. But to preserve consciousness and existential purpose, do we not ask ourselves the questions about remaining in this infinite loop, or do we eventually merge at AI intersection?

The Multimodal Leap

Picture this: A small drone fleet inspecting a construction site. Each drone sees in 3D pan-tilt-zoom high-resolution cameras. It listens for ongoing mechanical faults with sensors and human dialogues and provides feedback to the onsite manager in real time with day-to-day reporting. It allocates tasks, learns from every inspection with reinforcement feedback loops, and corrects and reiterates the same. We call these agentic, multimodal, self-improving systems.

A reality of specialized sectors in AI is reshaping the technological foundation with human-machine collaboration. In healthcare, we see the multimodal agentic revolution in terms of how patient medical history is interpreted and analyzed to predict outcomes. In turn, this offers real-time assistance to health professionals. Concurrently, patients receive updates on doctor's instructions and medications. Also, we see warehouse robots and self-driving cars function as decision-making, self-improving intelligences.

The incoming massive shift of smart machines that run with multimodal AI and agentic systems not only perceive reality as their surroundings but also operate autonomously. These are now navigating humanity closer to superintelligence. Vision, sound, language, and sensor data, all of these modalities are processed in a specialized encoder and then fusioned together to build a unified representation of a moment in time. This allows complex reasoning and correlations between fused modalities to make the output more human-like for us to understand. Integrated and diverse industrial data streams have enabled solutions for object recognition, predictive analysis, and pattern comprehension. When co-relating humans to machines, we hear, speak, feel and interpret our environments. That's multimodality. Similarly when machines combine vision, language, data, and memory, they gain a better grasp of their surrounding world.

Computer Vision: See as Humans

In the human visual system, the eye captures light and then converts it to electrical signals to gain a dynamic understanding of the captured visual. As an act of evolutionary refinement, the brain analyzes gestures, expressions, features, and postures to match with stored collections of memories in order to filter relevancy and predict movements.

Scientists worked decades to push the boundaries of machine vision for making it near accurate to the human vision system, but it relied heavily on handcrafted features and rule-based algorithms. From the breakthrough development of significant edge detection algorithms to the segmentation process for dividing image into regions, the challenges of image understanding were still with feature extraction.

The classification of objects became more prominent with object recognition, 3D reconstruction, support vector machines, and random forest algorithms. This major leap could become a missing link to extract features for face detection. With the rise of deep learning, multi-layered computational neural networks began to recognize complex patterns, objects, and faces. Large-scale labeled datasets, increased computational power with graphics processing units, and innovative algorithms made an immediate and profound impact. Semantic segmentation, 3D perception, and visual reasoning pushed boundaries to integrate natural language processing with computer vision. But a human-like computer vision performance still has ongoing challenges around data efficiency and interpretability that often require better datasets and integrating vision with cognitive potential. In untangling the mysteries of biological vision with the power of artificial intelligence, we are getting closer to human-like computer vision.

Agentic Systems

Agentic capabilities in any system unlock autonomy. Imagine an autonomous decision-making system capable of setting goals, separating them into manageable tasks, selecting appropriate tools, and executing actions while adapting to dynamic environments. The very core of the agentic system lies under the concept of agency, independently defining goals by proactively perceiving its environment in addition to executing actions from a feedback loop. It is intertwined with agentic and human collaboration. Reasoning capabilities, along with memory and planning in agents, are driven by large language models. Planning, decision, and learning feedback loops are the very architectural components of agents that can be compared to the human cognitive function. Effective agents prioritize goals, anticipate outcomes with predictive modeling, and modify the plans.

The further advancement towards multi-agentic systems has redefined the power of collaboration between humans and machines where AI agents that collaborate and coordinate can achieve complex goals. Take the examples of swarm specialized robots for logistic operations, multi-agent financial trading bots, and autonomous inspection robots. Their ultimate goal is not to replace humans but to enable collaboration between humans and AI by sharing context and intent, task delegation, and intervention when needed by ensuring ethical alignment.

Evolving into agentic systems, AI is now initiating many complex workflows through intelligent collaboration, autonomously and adaptively making decisions for a growing number of industries. The vibrant network of these agents gives them a rich sense with memory and self-improvement, which then becomes self-evolving systems. We call it proto-superintelligence.

Upward Trajectory of Scaling to Superintelligence

The uncharted territory of science fiction, what we might call "superintelligence" is a scaling of size, more trained data, and bigger AI models exhibiting emergent capabilities. Together, data, computer, and architectural innovations makes systems computationally feasible and algorithmically efficient. The future of superintelligence significantly impacts the GenAI platforms and products and is linked to scaling potential. As a cognitive engine capable of robust generalizations, it influences not only how we interact with technologies, but also it drastically solves unprecedented global challenges.

To build such generative AI products and platforms, it is required to transition towards advanced reasoning and self-improving capabilities and building proactive models of the agentic layer. The scaling to superintelligence represents a massive AI alignment problem for humanity's learning curve.

AI Integration, Challenges, and Inequality Factor

Integrated AI has transformed each sector with a rapid rate of innovation and challenges. Personalized treatment plans and patient insights became possible in the health industry. Vision-driven quality control made manufacturing capabilities more productive. AI agents were able to detect fraud and model risk scenarios. Retail, education, and creative industries were remodeled. R&D cycles such as engineering materials, vaccines, molecule discovery, and test simulations became solutions in our lifetime, reducing the time cycle to months and weeks. These are all products that ultimately self-evolved from user data and feedback. And ultimately, we came to see microeconomic shifts from tedious repetitive human labor to model training and computation costs. The organizations have adapted an autonomous route with minimal to no human interventions. Agentic scientific capabilities have made it possible to self-generate hypotheses faster than a team could.

Challenges we face are unparalleled in regard to the integration of AI. Poor-quality and biased data could lead to unreliable insights. Avoiding this demands continuous data governance and model evaluation. Data privacy and security is a critical part of AI models with manipulative algorithms. Many AI tools could affect the business operations and how an individual could see their future job roles. Regarding social and economic inequality, AI can create profound implications if the systems are not designed ethically and thoughtfully. Algorithmic bias could make a vulnerable shift in policies for hiring, credit score, geography, criminal justice, and healthcare sectors. Eroding public trusts, AI-powered surveillance could lead to privacy violations. Recommendation systems can create bias to limit knowledge and data.

Call to Agency, Stewardship, and Exploration

Addressing the challenges we face to integrate AI requires an unwavering, proactive, and collective approach to achieve responsible AI by mitigating bias, allowing humans to intervene in AI decision-making, and promoting techniques that implement training AI models without compromising the use of sensitive information or data.

To steward and explore the novel pinnacle of AI, we need to be good at developing adaptability by pivoting when required and learning to stack hybrid skills that combine domain expertise with AI fluency while equally focusing on emotional intelligence, complex judgment, negotiations, trust, and cultural insight. Having agentic literacy to design and supervise AI systems with governance and safety is very important.

Seeing AI as a force of good influence, an amplifier, and a collaborator requires not only technical expertise but leadership capabilities at exponential levels with a statute of morality. Good stewardship builds a system to emphasize human flourishing, thus raising global prosperity, transparency, safety, and inclusion. This should allow us to progress towards a future that benefits humanity and shapes the trajectories of how machines serve civilization.

About the Author

Niyati Prajapati was born and raised in India. After graduating with bachelor's and master's degrees and while working in corporate, she became a prominent figure at the absolute forefront of robotics and artificial intelligence. She's made a profound impact on the industry, through her groundbreaking work, seamlessly integrating cutting-edge AI research with real-world applications in autonomous systems. Her pioneering contributions to advance technical capabilities in the field of machine learning and computer vision has significantly shaped commercial landscapes and strategic trajectories. She has led impactful projects at a global organization, specializing in advancing the research of artificial intelligence projects and building generative AI tools, products, platforms, and intelligent agents to enhance business outcomes and optimize customer experiences. In addition to her professional work, she is committed to build an impactful AI community to lead the next generation youth. She continues to be an advocate, advisor, strategic growth AI consultant, and a driving force in the evolution of AI and robotics by inspiring innovation and fostering significant progress within global AI communities.

LinkedIn: https://www.linkedin.com/in/npcodes/

CHAPTER 28

HOW TO DESIGN GOVERNANCE THAT ENABLES, NOT BLOCKS

By Vasanthan Ramakrishnan
Entrepreneur, Legal & Compliance Expert and Angel Investor
Chicago, Illinois

> *The best way to predict the future is to create it.*
> —Peter Drucker

The AI Imperative vs. Corporate Guardrails

In an era defined by digital transformation, AI is no longer a distant promise as it has become a critical driver of strategic advantage. With only 20% of companies using AI in at least one business line, today over 78% of companies use it, according to a McKinsey Survey.[1] But across Fortune 100 and high-growth startup environments alike, executives encounter a paradox: The very companies that stand to benefit most from AI's capabilities often maintain policies that inhibit its use. According to Tech.co, over 68% of business leaders think employees should not use AI without a managers' permission, according to technology authority.[2]

Official guidelines may cite concerns over data leakage, regulatory compliance, or intellectual-property protection, yet employees routinely turn to third-party AI tools to accelerate decision-making, craft client deliverables, or streamline repetitive tasks. This disconnect creates a silent tug of war between innovation and control, one that raises fundamental questions about how organizations can harness AI without sacrificing security or governance.

The tension arises wherever the urge to experiment collides with the need for oversight. On one hand, executives recognize that AI can surface insights from complex datasets in minutes rather than weeks, personalize customer interactions at scale, and reduce time spent on low-value work. On the other hand, each use case involving corporate-owned data represents a potential breach of confidentiality, a debit against a company's most precious asset. Strict policies, drafted to guard against worst-case scenarios, often fail to reflect the nuanced realities of everyday work. As a result, employees ignore or bypass these rules, creating unregulated "shadow" practices that escape both visibility and audit.

Yet avoiding AI altogether is no longer an option. Competitors embed machine learning into product roadmaps; legacy incumbents retrain entire workforces to think with data. The faster organizations can absorb AI responsibly, the better they can navigate market disruptions, from emerging regulations on consumer privacy to shifting industry norms around automation. The challenge, then, is to transform policy from a reactive instrument of restriction into a proactive framework for safe experimentation. When governance and innovation co-exist, companies unlock a winning cycle where clearer rules foster trust in AI, which in turn drives broader adoption under controlled conditions.

This chapter presents a three-pillar model for companies and executives seeking to reconcile the AI imperative with corporate guardrails. First, it argues for a policy-first approach, where governance grows in line with real-world workflows. Next, it outlines strategies for developing executive AI fluency, practical, bite-sized learning that equips leaders to sponsor and guide AI initiatives. Finally, it describes metrics and feedback loops that measure both compliance and impact, ensuring that policies remain informed by actual outcomes. By weaving these pillars together, organizations can craft a governance environment that neither smothers creativity nor permits unchecked risk, but rather channels AI's potential into durable, responsible value.

Three Pillars for Responsible AI Adoption

In many organizations, the official stance on artificial intelligence remains cautious, if not prohibitive. A recent *Financial Times* investigation found that employees at a pharmaceutical firm routinely turned to ChatGPT for coding, even as company policy remained silent or outright forbade external AI tools, forcing managers into a reactive posture of monitoring and ad hoc enforcement rather than deliberate governance.[3] This disconnect is not isolated. A McKinsey Global Survey reports that only 21% of enterprises using generative AI have established formal policies governing its use.[4] Meanwhile, frontline workers deploy public-language models to draft reports, analyze data, and automate routine tasks, creating a shadow practice that eludes visibility and risks sensitive data exposure.

The stakes are high. Organizations that shy away from AI risk ceding ground to more agile competitors: McKinsey finds that business units deploying generative AI report revenue uplifts comparable to those achieved by traditional analytical AI, yet only a fraction of companies have the governance frameworks in place to scale these gains safely. In heavily regulated industries, finance, pharmaceuticals, and legal services, the potential for data leakage or inadvertent IP disclosures carries not only reputational cost but also regulatory penalties.

The challenge, therefore, is to shift from policies that merely forbid AI use to governance that regulates its usage. Drawing on best practices, such as AI sandboxes that embed ethics and compliance at the workflow level, organizations can create "living" policies that evolve alongside new use cases. When governance is treated as a collaborative exercise between compliance experts and business leaders, it becomes an instrument of controlled experimentation rather than a blunt instrument of restriction.

Pillar One: Policy-First Innovation

Before examining each mechanism in turn, it is useful to survey the following three core components that give policy its power to enable rather than obstruct:

Co-Creation, Not Top-Down Mandates

Policies imposed without input from end users often clash with real workflows, triggering covert "shadow" AI use and elevated risk. By involving business leaders and practitioners in design, governance becomes aligned with actual needs and earns the buy-in that makes compliance feasible.

Why It Matters and the Solution

Absent user engagement, a policy's legitimacy and practical value erode. McKinsey's survey of generative AI adopters found that organizations with cross-functional governance teams are twice as likely to report both high adoption rates and low compliance incidents, compared to those with siloed policy groups. Without participation from business units, companies forfeit both the operational insights and the cultural trust needed for responsible rollout.

Invite representatives from compliance, IT, legal, and key business functions into joint policy workshops. Begin by mapping critical workflows, e.g., how deal teams prepare client presentations or how R&D groups analyze proprietary data, and identify where AI could enhance or endanger processes. Facilitate rapid prototyping of policy language and technical guardrails; then iterate based on user feedback. By co-creating policy, organizations build frameworks that reflect real use cases, command genuine respect from users, and lay the groundwork for rapid iteration and in-tool enforcement.

Rapid Iteration and "Living" Governance

A static handbook cannot keep pace with weekly changes in model capabilities, deployment patterns, or threat vectors. Agile review processes, modeled on software sprints, allow organizations to pilot new uses, observe outcomes, and update rules continuously, as Johnson & Johnson did when it paired back a centralized AI board after finding only 10% to 15% of projects drove 80% of value.[5]

Policies drafted in isolation and left untouched quickly become obsolete. AI capabilities evolve on a monthly cadence, which means that new models, novel use cases, shifting threat vectors, and a static rulebook cannot keep pace. When governance fails to adapt, it either stifles legitimate experimentation or creates blind spots for emerging risks, forcing organizations into a cycle of ad hoc patches and one-off exceptions.

But keep in mind that generative AI models and tooling advance rapidly: What was safe yesterday may expose new vulnerabilities tomorrow. There is also some regulatory uncertainty, and some jurisdictions are still defining AI regulations, so internal rules must flex alongside external requirements. High-growth teams also cannot wait six months for a policy sign-off on each AI pilot; excessive delay erodes competitive advantage.

Rapid Iteration and 'Living' Governance

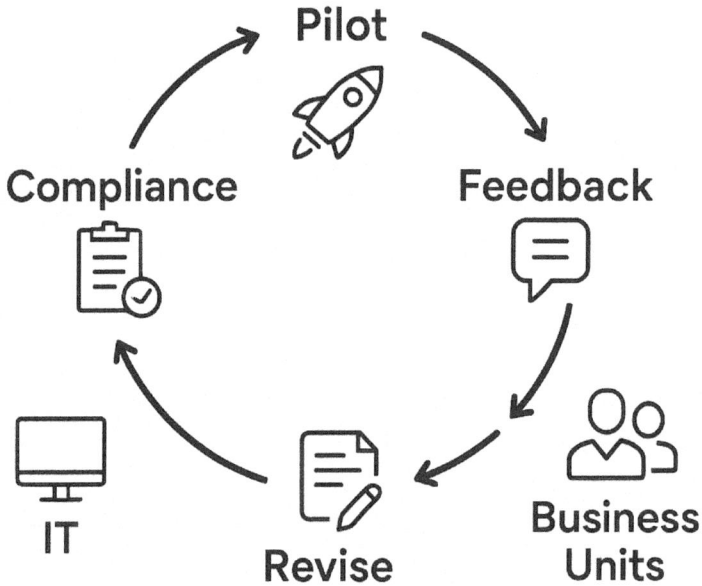

Figure 1 shows a living governance framework with a completed feedback loop.

The answer, hence, is an "agile solution."

1. *Time-Boxed Policy Sprints*

   - Establish a small steering committee of compliance, IT, and business-unit representatives.
   - Convene short, focused "policy sprints" (two to four weeks) to review specific use cases and draft controls,

and agree on interim guidelines.

- Iterate immediately after each sprint, publishing updates in a lightweight governance portal.

2. *Incremental, Use Case-Driven Updates*

- Apply a "5Ws" framework (who, what, when, where, why) to each new AI application: Assess risks and benefits case by case rather than expanding a universal policy.

- Adopt an incremental approach: Introduce guardrails for one domain (e.g., marketing copy generation), collect operational feedback, then extend or refine controls for subsequent domains.

3. *Continuous Monitoring and Feedback Loops*

- Instrument AI platforms with usage telemetry and policy-violation alerts.

- Review metrics, number of flagged incidents, time to remediate, and user feedback at biweekly governance check-ins.

- Deposit revised policy language and technical configurations into a shared "living" repository that tracks versions and rationales.

By treating governance as a living process, anchored in real use cases, advanced through discrete sprints, and grounded in continuous feedback, organizations can maintain rules that evolve as quickly as the AI tools they regulate. This dynamic model prevents policy from becoming either a bottleneck or a liability, ensuring that guardrails stay both relevant and robust.

Embed "Policy Hooks" in Tooling

Rather than relying on users to remember a separate policy document, companies can integrate automated prompts, pre-approved data scopes, and audit trails directly into AI platforms. Goldman Sachs's GS AI Assistant deployed to 10,000 developers behind the firm's secure internal platform, illustrates how built-in controls can both accelerate adoption and enforce compliance by design.[6]

Policies are only effective when they intersect seamlessly with daily work. Expecting users to consult a separate document before every prompt is unrealistic and invites shortcuts. Instead, governance must be woven into the very interfaces and platforms through which employees interact with AI.

Solution Patterns

1. *Automated prompts and warnings*: As soon as a user enters a prompt involving proprietary terms (e.g., product code names, client identifiers), the interface displays a contextual warning or requires explicit acknowledgement of policy compliance.

> **!** You're about to submit client data—
> please confirm compliance with
> Data Classification Policy

AI Chat

A   Sure, I can help with that.

Generate a report on ACME
Corporation's quarterly financals

Enter a prompt...

Submit

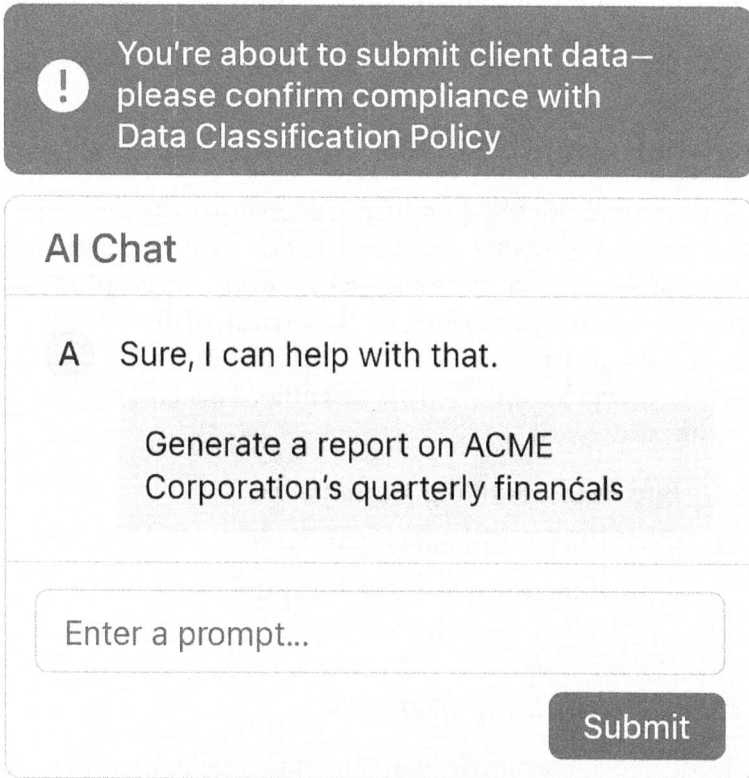

Figure 2 shows a financial-services firm's internal chat tool flags any prompt referencing customer account data and reminds employees of data-classification rules before submission.

2. *Scoped data connectors*: Rather than allowing free-form uploads, tools connect to approved data lakes or sanitized data views. Users select from pre-approved datasets, ensuring that only non-sensitive fields travel to the AI engine.

3. *Transparent audit trails*: Every prompt and response is logged in a secure ledger with metadata on user role, data

scope, and timestamp. Alerts trigger when unusual patterns emerge, such as high-volume exports or repeated attempts to bypass warnings.

Pillar Two: Executive AI Fluency

Organizations that fail to equip their leadership cadre with a practical understanding of artificial intelligence risk two equally damaging outcomes: either executives rubber-stamp every AI proposal without appreciating the attendant risks, or they reject worthwhile initiatives for fear of unknown liabilities. To bridge this gap, a structured approach to executive upskilling is essential. This pillar rests on three interlocking strategies:

1. Bite-sized upskilling journeys
2. Certification and badge programs
3. Cross-functional learning forums

1. Bite-Sized Upskilling Journeys

Senior leaders juggle strategic planning, stakeholder management, and P&L responsibilities. Lengthy training programs or academic courses seldom fit into their schedules, and highly technical sessions often fail to connect theory with the challenges of real deployments.

Why It Matters

Without targeted exposure, executives may rely on second-hand summaries or vendor pitches, leaving gaps in their judgment. Misaligned expectations can stall projects, lead to inappropriate risk tolerance, or allow unchecked experiments that run afoul of policy.

Solution Pattern

Micro-lessons: Short modules, 10 to 15 minutes each, focused on a single concept (e.g., "How generative models handle proprietary data" or "Assessing model bias in customer segmentation"). Content is delivered via on-demand video or interactive e-learners, allowing leaders to learn in between meetings.

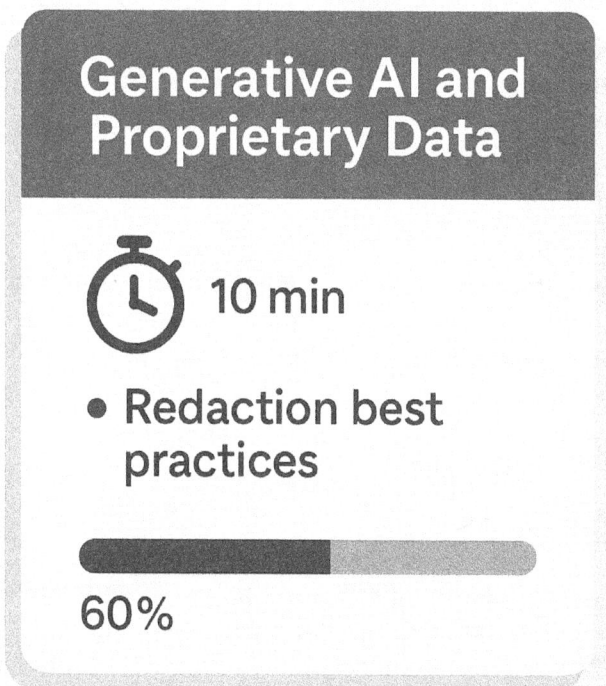

Figure 3 shows the progress from a sample micro-lesson on redaction best practices.

- *Just-in-time briefs*: "AI Office Hours" sessions scheduled twice monthly, where practitioners and compliance officers convene to discuss recent use cases, emerging risks, and regulatory updates. These drop-in clinics encourage live Q&A and build cross-functional trust.

- *Scenario-based simulations*: Lightweight, browser-based exercises that present a dilemma, such as choosing between an external LLM with advanced capabilities and an in-house model with stronger data protections, and require participants to weigh trade-offs under time pressure.

By meeting executives where they are, both in format and in relevance, bite-sized upskilling ensures leaders gain the insights they need without overwhelming their schedules. This foundation of practical familiarity lays the groundwork for informed sponsorship of AI initiatives.

2. Certification and Badge Programs

Informal recognition of AI competency, such as an offhand nod in a meeting, rarely translates into consistent stewardship. Without a structured credentialing process, executives and teams lack clear signals of readiness, making it difficult to allocate appropriate levels of access or responsibility.

Why It Matters

A formal certification program aligns incentives, rewarding participants for mastering privacy, security, and ethical considerations. It reduces the burden on compliance teams by creating a vetted pool of stakeholders who can advance to higher-risk projects with minimal oversight.

Certification and Badge Programs

 Bronze Access

Silver Access

Gold Access

 Certificate

Figure 4 shows a few sample badges that a participant can earn.

Solution Pattern

- *Tiered credentials*: Define clear requirements for each level (e.g., bronze for completing foundational modules, silver for passing a scenario-based assessment, gold for leading a cross-functional AI project).

- *Digital badges*: Issue verifiable badges that integrate with corporate directory profiles and project-management tools, signaling an individual's level of proficiency.

- *Renewal and recertification*: Set expiration dates and require short refreshers to maintain active status, ensuring skills stay current as models and policies evolve.

By formalizing certification and badge programs, organizations transform AI fluency from an abstract concept into a tangible asset, one that rewards competence, guides governance, and streamlines project approvals.

3. Cross-Functional Learning Forums

Knowledge silos undermine both innovation and governance. When data scientists, legal counsel, and business leaders work in isolation, misunderstandings proliferate: Engineers may underappreciate regulatory constraints, while lawyers may lack insight into technical capabilities. As a result, AI projects stall in the handoff between ideation and implementation, or they proceed without adequate risk assessment.

Figure 5 shows a cross-functional roundtable discussion involving legal teams and other teams.

Solution Pattern

- *Executive roundtables*: Quarterly sessions where senior leaders discuss AI initiatives, challenges, and trade-offs. Agendas rotate to broaden understanding.

- *Internal AI showcases*: Bi-monthly demos of prototypes and policy compliance dashboards. These forums promote accountability and highlight innovative governance.

- *Virtual knowledge hubs*: An intranet portal with case studies, policy updates, training, and a moderated discussion board, fostering active, community-driven learning.

Cross-functional learning forums knit together the disparate perspectives essential to responsible AI governance. By institutionalizing regular dialogue and knowledge sharing, organizations ensure that policy, practice, and technical expertise advance in lock step.

Pillar Three: Measuring Success and Feedback Loops

Before embedding policy into workflows and equipping leaders with practical AI knowledge, organizations must establish clear, empirically grounded measures of both risk and reward. Absent meaningful metrics, governance and training become untethered from real outcomes, policies may persist long after they cease to address pressing challenges, and upskilling efforts can proceed without evidence of impact. This pillar comprises three interrelated elements, each designed to translate abstract objectives into concrete indicators and to ensure continual refinement of both rules and practices:

1. Defining the right KPIs
2. Real-time dashboards and alerts
3. Structured feedback mechanisms

1. *Defining the Right KPIs*

Most enterprises default to vanity metrics, number of AI licenses issued, volume of prompts processed, or lines of code generated, to judge progress. Such measures say little about the balance of innovation and control. Also they do not capture the true value delivered or risks incurred. Without thoughtfully selected KPIs, teams cannot distinguish between safe experimentation and unchecked exposure, and executives lack the evidence needed to adjust policy or investment.

2. Real-Time Dashboards and Alerts

Even when teams agree on the right KPIs, manual reporting cycles, monthly or quarterly, fail to catch emerging risks in time to prevent data exposure. In rapidly evolving AI environments, a vulnerability exploited this week can escalate into a major incident before the next compliance review.

3. Structured Feedback Mechanisms

Dashboards and metrics yield quantitative insights, but they cannot explain why teams encounter friction or what contextual factors drive policy exceptions. Without qualitative feedback, governance updates risk overlooking root causes.

By defining precise indicators of success, surfacing real-time insights, and channeling user experiences into governance refinements, pillar three ensures that responsible AI adoption remains a data-driven, adaptive endeavor. Together with policy-first innovation and executive AI fluency, these measures complete a framework capable of steering organizations through the AI revolution without sacrificing security or control.

Roadmap for Integration

Translating the three pillars into sustained practice requires a phased approach. This roadmap guides organizations through a structured progression, from understanding current workflows and co-designing policy, to piloting training and measurement systems, and finally to broad deployment, automation, and continuous refinement.

- Phase 1: Discovery and policy co-design
- Phase 2: Pilot fluency programs and metrics baseline
- Phase 3: Scale, automate, and refine

Each phase builds on the last, ensuring that governance, education, and measurement mature in lockstep.

Phase 1: Discovery and Policy Co-Design

In this initial stage, the goal is to map existing processes, identify high-value AI use cases, and establish the first iteration of co-created policies. A few things that can help here are stakeholder workshops that convene small, cross-functional teams to document common workflows, such as report generation, data analysis, or client communications, and flag where AI could accelerate or complicate tasks. A risk assessment matrix for each identified use case can rate potential benefits against data-privacy, IP, and regulatory risks. And finally, in two- to three-week sprints, representatives from business units, IT security, legal, and compliance can draft policy language and tooling requirements in parallel.

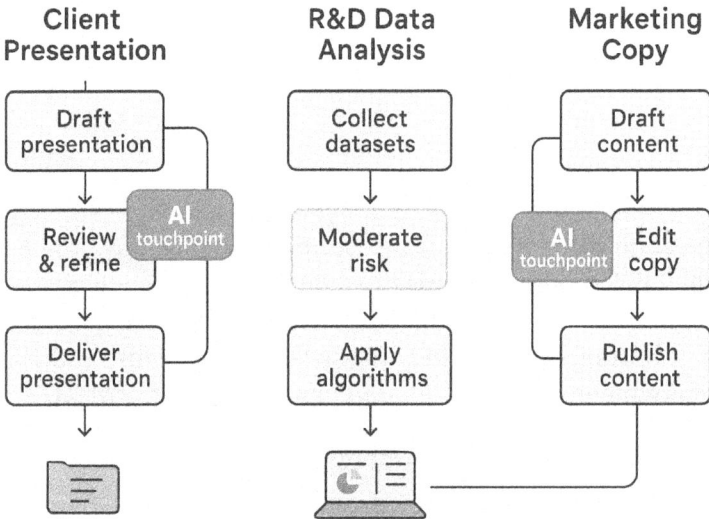

Figure 6 shows a sample workflow document with "AI touchpoints" where AI could be possibly used to add value.

Phase 2: Pilot Fluency Programs and Metrics Baseline

With policies drafted, the next step is to test both training and measurement on a limited scale.

- *Microlearning rollout*: Launch the first set of bite-sized modules and "AI Office Hours" for a pilot group of 20 to 30 leaders.

- *Badge program trial*: Offer bronze and silver badges to participants who complete foundational modules and pass a simple assessment.

- *Baseline metrics collection*: Instrument AI platforms to capture compliance events, prompt volumes, and time-savings data for the pilot cohort.

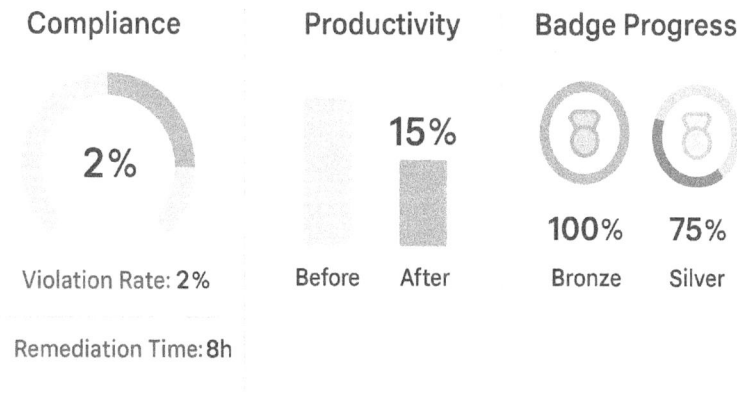

Compliance	Productivity	Badge Progress
2%	**15%**	**100%** **75%**
Violation Rate: 2%	Before After	Bronze Silver
Remediation Time: 8h		

Figure 7 shows a comprehensive dashboard that displays employee participation progress on the right with productivity change and the associated compliance violation on the left.

Phase 3: Scale, Automate, and Refine

Having validated both governance and training on a small scale, organizations can accelerate adoption across business units.

- *Automated policy hooks deployment*: Integrate reminders, connectors, and audit logging into enterprise-wide AI tools based on lessons from the pilot.

- *Full-spectrum badge certification*: Open gold-level credentialing for project leads who demonstrate advanced proficiency and have overseen compliant AI deployments.

- *Continuous improvement cadence*: Establish a quarterly review cycle in which the governance steering committee examines updated KPI trends, audit-trail summaries, and qualitative feedback; then prioritizes next steps.

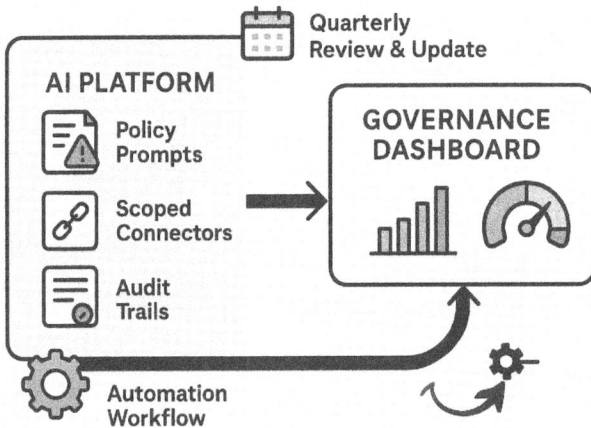

Figure 8 shows a continuous improvement framework involving a quarterly review cycle to periodically update the automation workflow of a company related to AI usage.

Conclusion

The pursuit of artificial intelligence adoption within large enterprises and high-growth startups invariably confronts a single, central challenge: bridging the divide between boundless technological potential and the imperative to safeguard sensitive assets. In this chapter, we introduced a three-pillar framework, (1) policy-first innovation, (2) executive AI fluency, and (3) measuring success with feedback loops, with each pillar addressing a critical dimension of responsible AI governance.

Ultimately, success demands leadership commitment, cross-disciplinary collaboration, and a willingness to view policy not as a barrier but as a strategic asset. By weaving together adaptive governance, leader-centric learning, and data-driven oversight, organizations can unlock AI's transformative promise while maintaining the integrity, privacy, and trust that underpin sustainable growth. The journey toward responsible AI is neither linear nor finite; rather, it is a perpetual cycle of innovation guided by the principles we have outlined, a cycle that equips enterprises to harness the full potential of artificial intelligence in a world where the only constant is change.

References

1. McKinsey & Company. *The State of AI in 2023: Generative AI's Breakout Year.* McKinsey Global Institute (McKinsey & Company), August 1, 2023. https://www.mckinsey.com/capabilities/quantumblack/our-insights/the-state-of-ai-in-2023-generative-ais-breakout-year.

2. Tech.co. "About Us." *Tech.co.* Accessed August 23, 2025. https://tech.co/about.

3. Stephanie Stacey. "Bosses Struggle to Police Workers' Use of AI." *Financial Times*, December 16, 2024. https://www.ft.com/content/cd08b45d-12dc-447e-bd59-1c366a7e6396.

4. McKinsey & Company. Eric Lamarre, Alex Singla, Alexander Sukharevsky, and Rodney Zemmel. *A Generative AI Reset:*

Rewiring to Turn Potential into Value in 2024. McKinsey Quarterly, March 4, 2024. https://www.mckinsey.com/capabilities/mckinsey-digital/our-insights/a-generative-ai-reset-rewiring-to-turn-potential-into-value-in-2024.

5. Isabelle Bousquette. "Johnson & Johnson Pivots Its AI Strategy." *Wall Street Journal*, April 18, 2025. https://www.wsj.com/articles/johnson-johnson-pivots-its-ai-strategy-a9d0631f.

6. Isabelle Bousquette. "Goldman Sachs Deploys Its First Generative AI Tool across the Firm." *Financial News London*, June 27, 2024. https://www.fnlondon.com/articles/goldman-sachs-deploys-its-first-generative-ai-tool-across-the-firm-1229ce44.

About the Author

Vasanthan Ramakrishnan is an entrepreneur, nonprofit leader, and bestselling author with a passion for empowering teams and redefining success. He is the founding partner and principal consultant at Ascend HSI Advisory Partners, an international legal services firm that works with Fortune 100, Fortune 500, and startup clients in areas including immigration, legal compliance, and international operations and compliance. As the youngest recipient of an honorary doctorate at 26, his journey spans award-winning initiatives like the Feminist Pen Foundation, where he led diverse teams across seven countries, to his current role as CEO of a thriving seven-figure business with a 50-person team operating in two countries. His upcoming book, *Breaking 9 to 5: Playbook for the EB-1A Einstein Visa (Free Resources Included)*, offers practical insights drawn from his extensive experience in nonprofit work and corporate leadership, inspiring others to unlock their full potential.

LinkedIn: https://www.linkedin.com/in/vasanthan-ramakrishnan/

CHAPTER 29

THE INTENTIONAL CYBORG: HUMAN UPGRADE REQUIRED

By Ruan Schutte
Innovation and AI Projects Specialist
Johannesburg, South Africa

If I had an hour to solve a problem, I'd spend 55 minutes thinking about the problem and five minutes thinking about solutions.

—Albert Einstein

The Frost Test

It was two degrees below zero Celsius. That's what I tell myself, anyway. The kind of cold that burns your nose when you breathe but sharpens your mind. I was out on a run with my wife—training for a new adventure, our first ultra-distance trail race. We were mid-run across the quiet golf course when I saw it: a patch of frost on the grass, glistening like snow under the first rays of the sun. Without thinking, I dropped onto the frozen turf and made a frost angel. No camera. No spectators. Just a grown man laughing in the frost, arms and legs flapping in childlike rebellion against adulthood and seriousness. My wife just shook her head and smiled. I always try to do something new or "something silly" on our outings. When was the last time your AI wanted to explore and challenge itself to something? AI would never have done that.

That moment stuck with me—not because of the cold, but because of the clarity. We're entering a world of intelligent machines—smart, fast, optimized. But none of them will ever stop mid-run just to make a frost angel. Not because it's efficient. Not because it has meaning. But because—why not? That's the part of us that doesn't get automated. That's the part that makes us human. That's the compass for becoming the kind of cyborg worth building.

We Are Already Cyborgs

You don't need implants or robot limbs or a USB port to be a form of cyborg, a soft cyborg. For the purposes of this chapter, we will just refer to "cyborgs" and not make a distinction between the two.

Every time you look at your phone to remember what you forgot. Every time Google Maps tells you where to go. Every time your smartwatch nudges you to go faster as you are not in the right heart rate zone, you're outsourcing part of your human experience or cognitive processing to machines.

But AI is different. It doesn't just enhance your memory or movement; it enhances your mind if used correctly. It plans, generates, rewrites, prioritizes, visualizes, and optimizes. It completes your thoughts. It even argues with you—sometimes better than your colleagues.

Unlike your calculator or calendar, AI talks back. This is not about smarter tools. It's about smarter selves. So the question isn't, "Should I become a cyborg?" It's, "What kind of cyborg am I becoming?" And more importantly, "Am I becoming one by design or just by default?" Each day, I work with tools, code with AI, write with AI, think with machines, google things, and plan according to my smartwatch. I'm not just writing about cyborgs; I am one.

Design Thinking—The Cyborg Upgrade Protocol

To become the best cyborg by design, then curiosity should be your fuel, and a human-centered thinking framework, like Design Thinking, should be your GPS. Design Thinking helps us keep humanity at the center of all innovations, especially when collaborating with artificial intelligence. During each stage, AI can serve as an amplifier:

- *Empathize*: AI can help you identify overlooked stakeholders, simulate diverse personas, generate rich backstories, and answer questions as if it were your customer.

- *Define*: It can apply frameworks like the 5 Whys or propose entirely new ways to explore your problem space, reframing challenges from multiple angles and using multiple thinking tools.

- *Ideate*: Use tools like brainstorming, future casting, intergalactic thinking, surreal art, and role play and let AI enhance and remix your ideas or inspire you to other ideas.

- *Prototype*: Create mockups, podcasts, images, and designs, generate test scenarios, even use AI to run simulations, and even create prototype applications with "vibe coding" now. You can build and iterate faster now than ever before.

- *Test*: AI can simulate user interactions, draft surveys, and complete surveys from different perspectives. It can analyze qualitative and quantitative feedback and data at scale.

The best AI collaborators won't be those who know the most about machine learning. They'll be those who understand people, problems, and possibilities—the human side of the equation.

The Cyborg Compass

Not all AI-enabled cyborgs are equal. Some let AI guide them entirely, hands off the wheel. Others refuse to use it "directly"; however, it's already embedded in many capabilities we use such as our phones, movie recommendations, GPS travel routes, online shopping product recommendations, etc.

But the next best cyborgs? They know when to ask for directions, when to wander, and when to set a new direction and change route and sometimes maybe just to see what's out there.

So, what helps us navigate this partnership? We need a compass, not just one that points north, but one that orients us toward meaningful collaboration between our human instincts and machine capabilities. Here's what that compass looks like:

- *Curiosity* that fuels exploration. You ask AI to simulate multiple customer types, map different onboarding journeys, and suggest radically different ideas from other industries.

- *Context* that grounds decisions. You use AI to analyze your company's operational data, understand constraints, and market opportunities, regulatory, or capacity challenges that shape feasibility.

- *Creativity* that generates new paths. You prompt AI to role play as the customer, highlighting pain points and emotional friction. You explore how this journey feels, not just how it performs.

- *Compassion* and empathy that reminds us of who we're solving problems for. You ask your AI what your customer, user, friend, partner, manager, etc., would like about the idea, what they would change, and what they would add.

- *Courage* that keeps us moving when the road disappears. You ask AI to critique your favorite ideas. You can even ask, "What would a future competitor do that we're too afraid to try?"

These are our coordinates. They remind us of what kind of cyborg we're becoming. They keep us from losing the "human" in human-machine.

I often tell people in my workshops: "AI isn't just a tool. It's a mirror and a canvas." Some of us unlock our best ideas with logic, steps, and structure. Others? We need rhythm, metaphor, a visual cue—or even a splash of chaos.

I once facilitated a multi-day workshop in Kenya with participants from across Africa. One of the activities I designed was intentionally unconventional. Each group was given a blank canvas and asked to paint their vision of Africa, blending in the theme of the event. But here was the twist: At the end, all the individual canvases had to connect, literally and thematically, to form one large piece of art that also looked like Africa.

Some teams collaborated intensely, negotiating where lines would meet. Others just painted joyfully, unconcerned with how their edges would align. One group made a color-coded spreadsheet to manage cross-team harmony. Another cranked up music, danced a bit, and painted freely.

Each team approached the same task with completely different thinking modes—analytical, intuitive, visual, playful. And when the canvases came together? It wasn't uniform. It wasn't "aligned." It was better! It was complex, vibrant, layered, and uniquely human. That's the point.

AI can help you brainstorm, generate, and remix. It can sketch and simulate and propose. But your thinking style—how you frame, explore, and make meaning—is still yours to cultivate. The best "prompts" are not just clever, they are human.

AI's multimodal capabilities means that any idea, problem or concept, no matter how complex or seemingly simple, can be transformed into an infographic, a poem, a surreal image, a short AI-generated podcast, a hypothetical debate between two personas, a roadmap with time and resource overlays, even a concept solution design or pitch, and many more.

Overlay these outputs across the compass. Does the image spark curiosity? Does the podcast dialogue build compassion? Does the chart sharpen your context? Does the story challenge your courage?

In my workshops, I encourage people to have AI ask them questions—for example, "What questions have I not asked about the idea, problem, or solution?"—or question the AI with, "What am I not seeing that could help us understand this better?" You could also be quite specific around the compass lenses, i.e.:

- "What are you still curious about that you haven't dared to ask?"
- "What context are you ignoring because it's inconvenient?"
- "What if a child or a philosopher approached this?"
- "Who will benefit most, and who might be left behind?"
- "What would courage look like that can be the practical next step?"

It's not always comfortable. But it shifts something. I recently asked AI to critique one of my concepts by giving me "negative social media comments" as if it were commenting after watching the YouTube version of a concept I was planning. The results were harsh but so informative. This is how we make the compass more than metaphor. We make it multimodal. Interactive. Provocative. Personal.

Why We Ask Boring Questions—And How We Could Fix It

Many people don't get to expand their own learning or thinking skills or gain insights from the experience of collaborating with AI. Not because the AI isn't good—but because the questions are boring. Chasing the answer, the destination is not nearly as rich or rewarding as the lessons, wonders, and experiences of the journey.

We ask safe, predictable, surface-level prompts: "Write me a strategy," "List 5 ideas," or "Summarize this." But AI tends to mirror our inputs. Garbage in, garbage out, just with perfect grammar and a "polite tone" (depending on the AI model you use). I would also caution here that some AIs are just too "nice" to us by default and design.

Here's the trick: If you want AI to be more interesting, you have to be more interesting first. If you want a more interesting life, you need to do more interesting things. You don't need to be a poet or a philosopher. You just need to unlock the same thing that makes humans compelling in conversations: intent, imagination, context, emotion, and—my personal favorite—exploration of new journeys and doing things you have not done or asked before.

I often remind my Design Thinking workshop participants: "Don't jump to solutioning mode." It's the most common pitfall I see, even among experienced professionals. We crave the answer, the shortcut, the certainty. The myelin "highways" in our brains want to shortcut to the "known" paths.

Now enter AI, our new cyborg upgrade: It's a destination rocket and autobahn for quick ways. It will happily skip the journey and take you straight to the next step. It suggests. It completes. It predicts. Just look at the suggestions after giving AI a prompt.

But here's the thing: Destination ≠ insight or growth. The beauty of Design Thinking is that it deliberately slows you down, helps you stay with the problem, explore the unknown, test assumptions, and evolve your thinking. When you pair AI with Design Thinking, and steer with your compass, you have a blueprint for the next great cyborg version. Here's how I encourage people to break out of default prompting:

- "What's the most counterintuitive solution to this problem?"
- "If this concept were a flavor, a scent, or a spoon, what would it be, and why?"
- "What would a 9-year-old do with this idea?"
- "What would a hostile boardroom, a curious monk, and an alien visitor say about this?"

- "What's the scariest assumption we're making without realizing it?"

AI is not here to give us answers we already know. It's here to stretch the way we think, if we stretch first. The Cyborg Compass and Design Thinking is your upgrade path. They make sure your journey with AI is human-led, not just machine-driven.

DiCon Dancing—Divergent and Convergent Thinking Modes for the Modern Cyborg

To become a better cyborg, you need to improve your thinking rhythm, a deliberate dance between imagination and focus. The real cyborg AI upgrade isn't hardware. It's rhythm. Design Thinking gives us a rhythm: the alternating current of divergent and convergent thinking, which I just call "DiCon dancing."

Divergent thinking is your expansion mode. You explore. Imagine. Suspend judgment. Push the boundaries. Act like a child without too many societal thinking constraints. Convergent thinking is your decision mode. You filter. Refine. Prioritize. Choose what matters. One big challenge is, most people don't know which mode they're in; they mix them at the wrong time. Or worse, they get stuck in one.

The best cyborgs learn to master both modes and switch between them on purpose. AI is an incredible thinking or DiCon dancing partner, but only if you're clear on your mode, your next move. In divergent mode, ask AI to stretch your imagination:

- "What's a solution no one would dare try?"
- "List 10 absurd but interesting ways to reframe this challenge."
- "If this was solved by a time-traveling artist, what would the idea look like?"

In convergent mode, use AI to sharpen your focus:

- "Compare these three ideas by desirability, viability, and feasibility with reasons."
- "Which of these options aligns best with long-term impact and why?"
- "Which decision-making framework can we use to make a final decision and explain how we will use it?"

AI can do both modes. But it mirrors your lead. If your thinking is unclear, AI will be, too. Many cyborgs today use AI like a search engine. The upgraded-version cyborgs, the DiCons, use it like a thinking or dance partner, with range and rhythm.

Cyborg Rights, Dependencies, and the Future of AI-Human Coexistence

We don't talk about it much, but maybe we should: Should every human have access to an AI as a cyborg right—like we have access to clean water, education, or basic healthcare? If the future belongs to AI-augmented minds, then access to intelligent tools becomes a form of cognitive inequality. Some people will have superpowers. Others will be left behind.

And it's not just about tools—it's about co-dependence. Not long ago, I injured my foot and had to use crutches for a while. Physically, I was temporarily incomplete—but still fully human. Then ChatGPT went offline for a few hours. My thinking partner was offline, mid projects. Maybe a "Cyborg Code" can help us stay intentional, not just intelligent:

- *Stay human on purpose*: Don't let speed replace soul or the journey. Make room for reflection, emotion, and fun—yes, even frost angels.

- *Be curious, not just correct*: AI can find some answers. You must find better questions. Wander off-script. Explore edge cases. Ask "what if?"

- *Retain the right to be wrong*: Failure is where humans learn the most. AI may avoid mistakes, but it also avoids surprise. Keep space for play.

- *Challenge the machine—and yourself*: Ask AI to critique you. Then critique its critique. Your job isn't to obey—it's to co-think and co-explore; like a Sherpa, you lead the vision.

- *Know when to disconnect*: Augmentation without pause becomes automation. Protect your mental sovereignty. Step away sometimes, go on a new adventure, do something you have not done before.

- *Share the power*: If AI makes you smarter, help others grow too. Democratize access. Teach the craft, not just the tool.

- *Keep the compass close*: Let curiosity, context, creativity, compassion, and courage guide your integration. Use tools to amplify purpose, not just productivity.

About the Author

Ruan Schutte is an explorer, husband, father, brother, and son, his greatest adventure and the most meaningful part of who he is. Ruan is an innovation specialist with a strong focus on applying artificial intelligence in both corporate and startup environments, across sectors ranging from agriculture to financial services. He leads strategic conversations on AI adoption and capability building and facilitates workshops that merge AI with design thinking and real-world problem-solving. A seasoned public speaker, Ruan frequently engages with thought leaders, tinkerers, and innovators across a range of forums—sharing ideas, sparking dialogue, and helping others navigate the evolving human-AI partnership.

Beyond boardrooms and whiteboards, Ruan is an adventurer at heart. He's completed the Iron Man, paddled the Duzi Canoe Marathon, tackled MTB and running trails, four-by-four overland routes in Southern Africa, and explored the depths through scuba diving. Recently, he's begun venturing into adventure motorbiking as his next frontier. This spirit of exploration shapes his work—he approaches innovation like an expedition: with curiosity, resilience, and creativity. In a world rushing forward, may we not forget that the most powerful upgrades still come from the human spirit—our willingness to explore, imagine, connect, and leave frost angels in the places no algorithm would ever look.

Email: Ruan.Schutte@outlook.com

LinkedIn: https://www.linkedin.com/in/idea8/

CHAPTER 30

HUMAN-CENTERED AUTOMATION: DESIGNING WITH CONSCIENCE, BUILDING WITH INTENTION

By Sheily Sharma
Architect, Intelligent Systems; Advocate, Ethical Automation
London, England, United Kingdom

> *To lead people, walk behind them.*
> —Laozi

Introduction: The Augmentation Mandate

The narrative surrounding artificial intelligence is often dystopian, predicting human obsolescence in a zero-sum game against machines. This perspective, however, overlooks a more powerful and valuable vision for our technological future: one of augmentation, not replacement. This is the foundation of human-centered automation (HCA), a paradigm that seeks not to diminish humanity but to amplify it. Human-centered automation is an approach to designing technology with the explicit goal of enhancing human capabilities. It reframes the relationship between people and machines from one of competition to one of collaboration.

Through this symbiotic partnership, the unique strengths of each are combined to achieve outcomes that neither could accomplish individually. The core value proposition is that the equation of "human + AI" yields a result far greater than either variable in isolation. In this model, technology becomes a "supertool," an extension of our intellect designed to make us smarter, faster, and more effective.

This vision relies on two core technical pillars: artificial intelligence and smart machines, offering analytical and predictive capabilities, and cloud-native DevOps, serving as the operational foundation that enables agility and automation in building and managing these sophisticated AI systems with a user-centric approach. This chapter will explore this emerging landscape, from its philosophical principles to its practical applications, and examine its profound impact on the future of work.

Redefining Automation: From Machine-Centric to Human-Centered

To appreciate HCA, one must understand the paradigm it evolves from. For decades, automation has been overwhelmingly machine-centric, prioritizing metrics of pure efficiency: maximizing output, reducing costs, and increasing speed. In this model, humans are often relegated to monitoring roles, expected to adapt to the rigid constraints of the technology.

Human-centered automation (HCA) represents an evolutionary shift, rooted in human-centered design (HCD), which places human needs and well-being at the core of the design process. Instead of asking, "How can we automate this human task?" HCA asks, "How can we use technology to help this human achieve their goals more effectively?" This reorients the purpose of automation from replacing labor to augmenting human intelligence.

Human-centered AI (HCA) is a framework that guarantees artificial intelligence serves humanity's best interests, guided by four core principles.

Principle 1: Augmentation, Not Replacement

The main goal of HCA is to enhance human intelligence, not to substitute it. AI should act as a powerful tool that boosts our cognitive abilities, fosters creativity, and supports complex decision-making. This approach positions AI as a partner that enables users to perform tasks more effectively, leading to outcomes superior to what either a human or a machine could achieve alone. The focus is on developing systems that improve, rather than replace, human skills.

Principle 2: Empathy and Ethical Design

At its heart, HCA requires that systems are built with empathy and a strong ethical basis. This involves a firm commitment to fairness, transparency, and accountability. Developers must proactively address potential biases in algorithms and prioritize user privacy and data security. By designing AI technologies that are safe, dependable, and respectful of human values and dignity, we ensure they benefit society fairly.

Principle 3: Collaborative Interaction

HCA considers the human-AI relationship as a dynamic, cooperative partnership. This interaction taps into the unique strengths of both: the machine's computational power and pattern recognition, combined with human intuition, common sense, and contextual understanding. The interface between them must be seamless and easy to use, allowing for smooth communication and ensuring the human user remains in control.

Principle 4: Focus on Meaningful Work

A key goal of HCA is to automate dull, repetitive, or dangerous tasks, freeing humans to focus on more meaningful work. By delegating routine tasks to machines, this framework enables people to focus on strategic thinking, creative problem-solving, and interpersonal collaboration—activities that require uniquely human skills and lead to greater personal and professional fulfillment.

Machine-centric automation and human-centered automation embody two radically different philosophies regarding the role of technology in our lives. The traditional machine-centric approach relentlessly aims to maximize efficiency and minimize operational costs by viewing human labor as expendable. Within this paradigm, technology operates as an autonomous substitute for human tasks, making data-driven, algorithmic decisions with little to no human intervention. Consequently, the role of humans is diminished to that of passive monitors, intervening only when a system fails. This model forces individuals to conform to the rigid processes dictated by machines, prioritizing scalability, reliability, and cost reduction over human needs.

In stark contrast, human-centered automation advocates for a powerful partnership between humans and technology. Its primary mission is to enhance human capabilities and make work not just productive, but meaningful. Here, technology is intentionally designed as a tool that supports and amplifies human intelligence, empowering individuals to embrace more strategic and creative roles. They can define goals, curate processes, and guide outcomes, with decision-making evolving into a collaborative effort that requires essential human insight. This people-first philosophy elevates user well-being and job satisfaction, insisting that technology must be adaptable and responsive to human needs, ensuring that progress truly serves people first.

This human-centric approach is not just about well-being; it's a superior business strategy. Machine-centric systems are efficient but fragile, struggling with volatility. By keeping humans "in the loop," HCA leverages our unique adaptability and creative problem-solving, creating systems that are inherently more resilient. Designing for people is a direct investment in a more robust and agile organization.

The Engine Room: Cloud-Native DevOps and the Reduction of Toil

The philosophy of HCA resonates powerfully within the realm of cloud-native DevOps. At its heart, DevOps represents a transformative cultural movement designed to confront and dismantle the friction and inefficiencies that often plague the interactions between development (dev) and operations (ops) teams. By fostering a spirit of collaboration and a sense of shared ownership, DevOps creates a harmonious and invigorating environment for technology professionals, turning the workplace into a dynamic hub of creativity and innovation.

A cornerstone of the DevOps methodology is the continuous integration/continuous deployment (CI/CD) pipeline, an essential mechanism that automates the laborious and error-prone tasks typically associated with software releases. This automation dramatically reduces cognitive load—the mental effort and strain required to execute complex tasks—while also eliminating toil, which refers to the tedious and repetitive work that offers little to no value in the grand scheme of software development. With these burdens lifted, developers are empowered to channel their energy into more creative and high-impact projects, ultimately making their work not only faster and safer but also profoundly more meaningful and fulfilling.

As systems grow more complex, the practice of observability—instrumenting systems to emit rich telemetry data (logs, metrics, and traces)—becomes crucial. This provides deep visibility, allowing teams to understand not just that a problem occurred, but why. It transforms troubleshooting from a high-stress event into a structured investigation, reducing frustration and improving system reliability.

A mature DevOps culture, with its automated pipelines and robust observability, is a prerequisite for deploying responsible AI at scale. The operational agility of DevOps provides the framework needed to deploy, monitor, update, and, if necessary, roll back AI models quickly and safely, turning the principle of "responsible AI" into an operational reality.

The Symbiotic Loop: AI-Powered DevOps (AIOps)

Just as DevOps lays the groundwork for human-centered automation (HCA), AI now seamlessly integrates into the operational process, forging a powerful, symbiotic alliance known as AIOps (artificial intelligence for IT operations). Leveraging advanced machine learning, AIOps revolutionizes IT management by automating tasks and tackling the complexity of modern architectures head-on. By scrutinizing immense volumes of operational data, AIOps systems swiftly detect anomalies and forecast potential issues before they trigger outages. This shift transforms the chaos of firefighting into strategic, proactive management, empowering teams to drive innovation instead of constantly reacting to crises.

AIOps shifts IT teams from a reactive to a predictive posture. By analyzing vast amounts of operational data, AIOps systems can detect anomalies and predict impending issues before they escalate into outages. This transforms stressful "firefighting" into proactive management, enabling teams to focus on innovation rather than crisis response.

In software development, AI tools now function as a "copilot" for developers, suggesting code, generating tests, and writing documentation. By automating the "grunt work" of coding, these tools increase productivity and reduce cognitive load, freeing developers to concentrate on complex architecture and creative problem-solving. A virtuous cycle emerges: Human-centric DevOps practices enable complex systems, and the resulting complexity is then managed by AIOps, which in turn makes the work of human operators even more human-centric.

Human-Centered Automation in Practice

The impact of HCA becomes tangible when examined through real-world applications where technology augments, rather than replaces, human judgment.

On the Factory Floor

- *Collaborative robots (cobots)*: Designed to work alongside people, cobots handle physically demanding tasks, such as lifting heavy parts, while human partners perform delicate assembly work. This reduces the risk of common manufacturing injuries and physical strain.

- *AI-powered quality control*: AI vision systems can detect common manufacturing defects with superhuman accuracy, flagging only novel or unusual issues for review by a human expert. This frees skilled workers for higher-level analysis and process improvement.

- *Predictive maintenance*: AI systems analyze data to predict when machinery is likely to fail, allowing for proactive maintenance that prevents dangerous and costly breakdowns.

In the Hospital Ward

- *Clinical decision support*: AI algorithms act as a "second pair of eyes," analyzing medical images or patient data to highlight subtle anomalies for clinicians. The final diagnosis and patient conversation remain firmly in the hands of the human healthcare professional, such as a doctor or nurse.

- *Reducing administrative burnout*: Ambient AI tools listen to doctor–patient conversations and automatically draft clinical notes for the physician to review and approve.

This frees clinicians from burdensome data entry, allowing them to focus their full attention on the patient.

In the Creative and Knowledge Professions

- *AI as a creative partner*: For artists and writers, generative AI can act as a brainstorming partner, generating initial concepts or drafts to accelerate the creative process. The human remains the essential curator and final artist.

- *The augmented analyst*: In fields like finance and law, AI automates the laborious process of data collection and analysis, allowing human experts to focus on higher-value work like interpreting insights, developing strategy, and communicating findings

Across these examples, a clear principle emerges: The system delegates tasks of scale to the machine, while reserving tasks of sense-making for the human.

The Evolving Workforce: New Skills for More Meaningful Work

HCA does not lead to a future without jobs; it prompts a re-evaluation of the tasks within jobs. It automates tasks, not entire jobs, leading to the "great task-shift." A customer service agent, for example, may be freed from routine inquiries by a chatbot, allowing them to focus on high-skill problem-solving and address complex and emotionally charged issues.

As AI handles routine tasks, the economic value of uniquely human competencies will soar. Skills such as critical and creative thinking, emotional intelligence, empathy, collaboration, and communication become the essential differentiators of an augmented workforce, as these are abilities that machines cannot easily replicate. Alongside these, new technical skills, such as AI and data literacy, will become essential for effectively leveraging these new tools. Many roles will evolve toward orchestrating AI systems—setting goals, handling exceptions, and ensuring alignment with strategic objectives.

This shift has the potential to increase job satisfaction significantly. By automating the most tedious and stressful aspects of a job, HCA enables employees to focus on more engaging, strategic, and creative work, thereby fostering a greater sense of purpose and combating burnout. This transformation represents a potential re-humanization of work, creating economic incentives for the very skills—creativity, critical thinking, empathy—that were often suppressed in the name of industrial efficiency.

The Essential Guardrails: Building Responsible and Trustworthy Systems

The vision of HCA is built on a foundation of trust, which must be earned and maintained. Responsible AI is not an optional add-on; it is a non-negotiable prerequisite for the entire paradigm to succeed. Key principles include:

- *Fairness and bias mitigation*: AI systems must be designed to provide equitable outcomes and avoid amplifying harmful societal biases.

- *Transparency and explainability (XAI)*: AI decision-making processes must be understandable to the humans who collaborate with and oversee them.

- *Accountability and governance*: Clear lines of responsibility must be established for the outcomes of AI systems.

- *Privacy and security*: Robust protocols for protecting data privacy and securing systems against malicious use are critical.

Algorithmic bias is a primary threat. For instance, AI hiring tools have automated gender discrimination by penalizing resumes with words associated with women, and facial recognition systems have shown higher error rates for women and people of color. Addressing this requires curating diverse datasets, conducting regular audits, and maintaining meaningful human oversight.

Upholding these principles requires a robust technical framework. This is the role of MLOps (machine learning operations), which applies DevOps principles to the machine learning lifecycle. MLOps provides the capabilities to monitor models for drift or bias continuously, trace data lineage for auditing purposes, and rapidly deploy updated models when issues are detected. The operational agility of DevOps and MLOps provides the necessary infrastructure to enforce the ethical guardrails of responsible AI. Investing in a responsible AI framework is not a constraint on progress; it is a direct investment in product quality, user trust, and long-term market acceptance, accelerating the deployment of truly human-centered AI.

Conclusion: The Future Is a Collaboration

Human-centered automation offers a compelling vision for the future. It requires a philosophical choice: to view technology as a tool for augmentation, not replacement. This vision is realized through the convergence of AI and cloud-native DevOps, all built on the essential guardrails of responsible design. The future of work is one of human-AI symbiosis, where technology handles scale and humans provide sense-making, elevating our roles to be more strategic and creative.

The most significant barrier is not technological but psychological. We must shift from a mindset of fear to one of collaboration and cooperation. By designing AI as a partner, we can unlock a new era of human potential, making work safer, faster, and profoundly more meaningful.

About the Author

Sheily Sharma has developed a curiosity about the technological evolution, particularly where emerging complex systems intersect with humans. She has worked on creating a box with no boundaries, combining DevOps and AI automation, which is crucial in a world where technology is not merely a tool of acceleration, but an instrument of ethical augmentation. With a high command of cloud-native architectures, MLOps, and autonomous systems, Sheily has created an ecosystem that harmonizes scalability with an understanding of societal responsibility. She has learnt about functioning in a manner where the contributions are a merging of a new technological ethos, one that dismantles the binary between human intellect and machine efficiency. Her thought leadership challenges societal norms in a way that promotes progress more safely and effectively.

Sheily advocates for multifaceted innovation, embedding empathy within algorithms and engineering inclusivity into scalable systems. Her vision is not of automation for convenience, but of automation for human-based needs, rather than replacement, ushering in a renaissance approach to the next era of digital transformation. The current decade, which is undergoing drastic changes with age, is ultimately defined by the mechanistic optimization of technologies to benefit humanity. Sheily wants to guide things in a manner of reclaiming humanity's central role in technological progress. Her trajectory is not one of mere contribution, but of catalysis—shaping the narrative of how intelligent systems evolve with, and for, human potential.

Email: slaywithsheily@gmail.com

LinkedIn: https://www.linkedin.com/in/sheily-sharma-8036641b2

CHAPTER 31

AI IN ROBOTICS: REINFORCEMENT LEARNING FOR ROBOTIC CONTROL

By Sakina Syed, B.Sc.
Microsoft IT Professional, AI Engineer and Consultant
Toronto, Canada

Artificial intelligence is growing up fast, as are robots whose facial expressions can elicit empathy and make your mirror neurons quiver.

—Diane Ackerman

Imagine a robot that learns to walk, grasp objects, or navigate complex environments—not by being programmed step by step, but by trial, error, and experience. This is the promise of reinforcement learning (RL) in robotics: empowering machines to develop intelligent behaviors through interaction with the world. As robotics ventures into increasingly dynamic and unpredictable domains, RL offers a powerful framework for autonomous decision-making, enabling robots to adapt, optimize, and thrive in real time.

Now, imagine a future where a humanoid robot, freshly deployed on Mars, learns to traverse alien terrain, build shelters, and repair equipment—all without prior knowledge of the environment. It trains in simulation, adapts through RL, and evolves its behavior with each challenge it faces. No manual programming, no remote control—just pure learning. This vision is rapidly becoming reality as RL reshapes how robots learn and evolve, pushing the boundaries of what machines can achieve without explicit instruction.

From industrial automation to household assistants and autonomous vehicles, RL is revolutionizing robotic control. This chapter explores the synergy between RL and robotics, diving into the algorithms, challenges, and breakthroughs that are driving the next generation of intelligent machines.

Fundamentals of Reinforcement Learning (RL)

Reinforcement learning (RL) is a foundational paradigm in machine learning that enables agents to learn optimal decision-making strategies through trial-and-error interactions with their environment. Unlike supervised learning, which depends on labeled datasets and predefined outputs, RL thrives in scenarios where feedback is sparse and indirect. The agent—often a robot or software entity— observes the state of its environment, selects an action, and receives a reward that quantifies the success or failure of that action. This reward signal acts as a compass, guiding the agent toward behaviors that yield higher cumulative returns over time.

At the heart of RL lies the concept of a policy, a mapping from states to actions that the agent or robot refines through experience. Supporting this is the value function, which estimates the long-term benefit of being in a particular state or taking a specific action. This allows the robot to weigh immediate rewards against future gains, fostering strategic and adaptive behavior. The learning process is inherently dynamic: As the robot explores, it continuously updates its understanding of the environment, improving its policy to better navigate uncertainty and complexity.

In the context of robotics, RL unlocks the potential for machines to autonomously master intricate tasks—such as balancing on uneven terrain, manipulating delicate objects, or coordinating multi-joint movements—without explicit programming. This ability to learn from interaction makes RL a powerful tool for developing intelligent, flexible, and resilient robotic systems capable of operating in real-world environments where unpredictability is the norm.

RL Algorithms in Robotics

Applying reinforcement learning (RL) to robotics introduces a distinct set of challenges that set it apart from other areas of machine learning. Robots operate in high-dimensional state spaces, often relying on rich sensory inputs such as vision, touch, and proprioception. Their actions are typically continuous, requiring precise motor control rather than simple, discrete decisions. Additionally, training in the real world demands sample-efficient learning due to safety risks, hardware wear, and time constraints. These complexities have led to the development and adaptation of specialized RL algorithms tailored to the unique demands of robotic control.

One of the earliest RL algorithms, Q-learning, for example, is well-suited for environments with discrete actions. However, its scalability is limited in robotics, where tasks often involve continuous control and high-dimensional inputs. To address this, Deep Q Networks (DQN) combine Q-learning with deep neural networks, enabling agents or robots to learn directly from raw sensory data like images. For example, DQN has been used in robotic arms to identify and grasp objects in cluttered scenes, leveraging visual input to make decisions without manual programming.

For tasks requiring smooth and continuous control—such as reaching, pushing, or locomotion—algorithms like Deep Deterministic Policy Gradient (DDPG) are more appropriate. DDPG uses an actor-critic architecture, where the actor suggests actions and the critic evaluates them. This approach has been successfully applied in simulation environments like OpenAI Gym's Fetch robot, which learns to perform pick-and-place operations with high precision.

Building on the strengths of DDPG, Proximal Policy Optimization (PPO) introduces a more stable and efficient learning process by limiting the extent of policy updates. This makes PPO particularly effective in real-world applications where stability is crucial. A notable example is Boston Dynamics' Spot, where PPO has been used in research settings to train the robot to walk, balance, and recover from disturbances across varied terrains.

To further enhance exploration and robustness, Soft Actor-Critic (SAC) incorporates entropy maximization into the learning process. This encourages agents to explore diverse strategies and avoid settling prematurely on suboptimal behaviors. SAC has been applied to manipulation tasks involving deformable or unpredictable objects, such as those handled by the Franka Emika Panda robot, which learns to interact with soft materials and dynamic environments using visual and tactile feedback.

Finally, Twin Delayed DDPG (TD3) improves upon DDPG by addressing overestimation bias through the use of twin critics and delayed updates. This results in more stable and accurate learning, especially in dynamic tasks like jumping, running, or precision assembly. TD3 is commonly used in simulation platforms such as NVIDIA Isaac Gym, where robots are trained in parallel across thousands of environments before transferring their learned policies to real-world hardware.

These algorithms are often integrated with complementary AI techniques—such as computer vision, simulation-based training, and transfer learning—to build robust and intelligent robotic systems. Whether the task involves navigating complex terrain, manipulating delicate objects, or performing industrial assembly, RL algorithms provide the adaptive learning capabilities that underpin modern autonomous robotics.

Sim-to-Real Transfer

Training robots directly in the real world is often impractical due to a combination of safety risks, high operational costs, and the time-intensive nature of physical experimentation. Robots can break, environments can be difficult to reset, and collecting enough data for learning can take days or even weeks. To overcome these limitations, researchers and engineers increasingly rely on simulation environments virtual platforms that provide a controlled, scalable, and risk-free space for training robotic agents.

Simulators offer several key advantages. They allow for fast, parallelized learning, where thousands of training episodes can be run simultaneously. Environments can be easily reset or modified, enabling rapid experimentation with different scenarios, tasks, and robot configurations. Most importantly, simulation eliminates the risk of hardware damage, making it ideal for early-stage development and testing of complex behaviors.

However, transferring a policy trained in simulation to a real-world robot introduces what is known as the reality gap—the discrepancy between simulated and real-world physics, sensor noise, and environmental dynamics. Even small differences in friction, lighting, or object mass can cause a policy that performs well in simulation to fail in the real world. To bridge this gap, several techniques have been developed:

- *Domain randomization* involves intentionally varying simulation parameters—such as textures, lighting, object shapes, and physical properties—during training. This forces the agent to learn policies that are robust to a wide range of conditions, increasing the likelihood of successful transfer to the real world.

- *System identification* takes a more precise approach by calibrating the simulator to closely match the real-world robot's dynamics. This involves collecting real-world data and tuning the simulation parameters—such as joint friction, mass distribution, and actuator delays—so that the simulated robot behaves as similarly as possible to its physical counterpart.

- *Fine-tuning* is often used after initial deployment. Once a policy is transferred to the real robot, it can be further refined using real-world data. This helps the agent adapt to subtle differences that weren't captured in simulation and improve performance over time.

Several powerful simulation platforms support these workflows, including Gazebo, MuJoCo, Isaac Gym, and PyBullet. These tools offer physics engines, 3D rendering, and integration with RL libraries, making them essential for modern robotics research and development.

Together, these tools and techniques make sim-to-real transfer a cornerstone of reinforcement learning in robotics. By enabling safe, scalable, and efficient training, simulation accelerates the development of intelligent robotic systems that can operate reliably in the unpredictable conditions of the real world.

Smarter Together: Unlocking Robotic Potential with Multi-Agent Reinforcement Learning (MARL)

Multi-agent reinforcement learning (MARL) extends traditional RL to environments where multiple agents interact—cooperatively, competitively, or in mixed settings. Unlike single-agent RL, each agent in MARL learns its own policy based on its observations, actions, and rewards, while simultaneously adapting to the behaviors of other agents. This interdependence introduces additional complexity, as the environment becomes non-stationary and evolves dynamically with each agent's learning process.

Learning to Cooperate and Compete: MARL's Role in Next-Gen Robotics

MARL is particularly valuable in robotics for tasks that require coordination, resource sharing, and distributed decision-making. In scenarios like autonomous vehicle fleets, warehouse logistics, or drone swarms, agents must not only optimize their own actions but also anticipate and respond to the actions of others. This enables more efficient, scalable, and intelligent multi-robot systems.

Collective Minds: How MARL Is Revolutionizing Robotic Coordination

In multi-agent reinforcement learning (MARL), agent interactions can be categorized into three main types: cooperative, competitive, and mixed. Cooperative interactions involve agents working together toward a shared objective, often requiring synchronized strategies and communication. A prime example is drone swarms collaboratively mapping terrain, where each drone must coordinate its flight path and data collection to efficiently cover the area without redundancy. Such cooperation is also evident in warehouse automation, where multiple robots jointly manage inventory tasks, optimizing for speed and accuracy.

Competitive interactions, on the other hand, arise when agents have conflicting goals and must outmaneuver one another to succeed. This dynamic is exemplified by robotic soccer teams vying for control of the ball, where each agent must anticipate opponents' moves and adapt its strategy in real time. Similar competitive behavior is seen in algorithmic trading bots competing for profitable market positions. Mixed interactions blend elements of both cooperation and competition, creating complex environments where agents must navigate shifting alliances and trade-offs.

Autonomous traffic systems illustrate this well: Vehicles aim to reach their destinations efficiently (individual goals) while also cooperating to prevent congestion and accidents (collective goals). Another example is multiplayer online games, where players may form temporary alliances to overcome challenges but ultimately compete for victory. These interaction types highlight the diverse and nuanced nature of MARL environments, demanding sophisticated learning algorithms capable of adapting to dynamic social contexts.

From Solo Bots to Swarms: The Rise of Collaborative Intelligence in Robotics

To support these interactions, MARL employs several key techniques. Centralized Training with Decentralized Execution (CTDE) allows agents to be trained using shared global information while enabling them to operate independently during deployment, enhancing scalability and robustness. Communication learning enables agents to determine when and how to exchange information, fostering improved coordination and situational awareness. Additionally, reward shaping and credit assignment techniques help agents discern their individual contributions to team success, promoting more efficient learning and fairness across the system.

Real-World Applications of MARL in Robotics

Application Area	Description	Example Algorithms
Autonomous Vehicles	Vehicles coordinate to merge, avoid collisions, and optimize traffic flow	MADDPG, QMIX
Drone Swarms	Drones collaborate for search-and-rescue, terrain mapping, or surveillance	PPO, COMA
Warehouse Robots	Robots share space and tasks efficiently in logistics environments	DDPG, VDN
Multi-Robot Exploration	Robots divide and conquer unknown terrain, sharing discoveries in real time	SAC, MAPPO

Bridging RL with Broader AI Paradigms

The convergence of RL with foundation models, natural language understanding, and world modeling is poised to revolutionize how robots perceive, reason, and interact. Imagine robots that not only learn from their environment but also understand human instructions, anticipate outcomes, and plan actions with foresight—blurring the line between reactive agents and cognitive collaborators. For instance, a household robot could interpret a spoken request like "set the table for dinner," infer the context (number of people, time of day, cultural norms), navigate the space efficiently, and execute the task with minimal supervision—all while adapting to changes such as unexpected guests or missing utensils.

Case Studies: Adaptive Learning in Robotics Through Reinforcement Learning

Reinforcement learning (RL) has emerged as a transformative paradigm in robotics, enabling machines to learn complex behaviors through trial and error, guided by feedback from their environment. From precision manipulation to agile locomotion and autonomous decision-making, RL has demonstrated remarkable success across a wide range of robotic applications. The following case studies illustrate how RL, often integrated with complementary AI technologies, is solving real-world challenges and driving the next generation of intelligent, adaptive robotic systems.

Surgical Robotics: Learning from Demonstration

In precision-critical environments like surgery, transferring nuanced human motion skills to robots poses a significant challenge. Learning from Demonstration (LfD) offers a compelling solution by allowing robots to observe and mimic expert surgical behavior, effectively bridging the gap between human intuition and robotic execution. This approach employs nonlinear dynamic systems to encode movement primitives, enabling adaptive execution of surgical tasks.

Recent advancements have introduced adaptive learning frameworks that incorporate real-time feedback mechanisms, such as haptic cues and stylistic behavior analysis, to refine robotic performance based on human-like movement patterns. For example, force feedback models—including spring-based, damping, and hybrid systems—have been employed to enhance motion fluidity and targeting precision during simulated surgical procedures.

A notable example is the da Vinci Surgical System, which utilizes LfD-based techniques and haptic feedback to assist surgeons in performing minimally invasive procedures with enhanced precision and control. These systems not only enhance dexterity but also personalize training by detecting deficiencies in movement style and adjusting guidance accordingly. As a result, surgical robots can tailor their motions to patient-specific anatomical conditions, significantly improving safety, flexibility, and effectiveness in clinical settings.

Locomotion: Agile Movement in Unpredictable Terrain

Legged robots face the challenge of maintaining balance, coordinating joints, and adapting to unpredictable terrain. RL has enabled breakthroughs in dynamic gait learning, recovery strategies, and terrain-aware movement. MIT's Mini Cheetah, for example, used deep RL and proprioceptive feedback to master walking, running, and recovering from falls—skills essential for navigating disaster zones and rough terrain. Similarly, Boston Dynamics' Spot, as also mentioned above, has leveraged RL to enhance its locomotion capabilities, allowing it to traverse stairs, slopes, and uneven surfaces with agility and resilience. These examples highlight RL's potential to create robust, adaptive locomotion strategies for real-world deployment.

Robotic Arms: Precision Manipulation and Dexterity

Robotic arms are central to industrial automation, logistics, and research, performing tasks such as grasping, pick-and-place operations, and intricate assembly. RL algorithms like Deep Deterministic Policy

Gradient (DDPG) and Proximal Policy Optimization (PPO) have proven effective in learning fine-grained control policies. A standout example is OpenAI's Dactyl—a robotic hand trained entirely in simulation using domain randomization to manipulate a Rubik's Cube using only vision and touch. Once deployed, the hand demonstrated impressive dexterity and robustness, solving the cube in the physical world without additional programming. This case exemplifies how RL, combined with sim-to-real transfer and computer vision, can enable complex, human-like manipulation skills.

Manipulation Tasks: Context-Aware Interaction

Beyond rigid object handling, RL is increasingly applied to complex manipulation tasks involving tool use, object stacking, and interaction with deformable materials like cloth or cables. These scenarios require real-time perception, decision-making, and adaptability. By integrating vision-based inputs with RL, robots can operate in cluttered and dynamic environments. For instance, the Franka Emika Panda robot was trained using Soft Actor-Critic (SAC) to manipulate soft objects, adapting its grip and motion based on visual and tactile cues. Such tasks showcase RL's ability to enable nuanced, context-aware manipulation in real-world settings.

Aerial Drones: Autonomous Flight in Dynamic Environments

Aerial drones face unique challenges in 3D navigation, obstacle avoidance, and real-time decision-making under uncertainty. RL has proven powerful for enabling drones to learn complex flight behaviors, especially when combined with simulation and vision-based perception. Researchers have trained quadrotors using PPO and SAC to autonomously navigate cluttered indoor environments, perform agile maneuvers, and land on moving platforms. For example, a drone developed at the University of Zurich could fly through a forest at high speed using only onboard cameras and a policy trained with RL and computer vision. These applications demonstrate RL's capacity to empower drones for tasks like search and rescue, package delivery, and environmental monitoring.

Autonomous Vehicles: Intelligent Driving in Complex Traffic

Autonomous vehicles (AVs) represent one of the most high-stakes applications of RL, requiring real-time decision-making in dynamic environments. RL has been used to train AVs for lane keeping, adaptive cruise control, obstacle avoidance, and multi-agent coordination. In simulation platforms like CARLA and AirSim, algorithms such as Deep Q Networks (DQN) and PPO have enabled vehicles to navigate urban landscapes, respond to traffic signals, and avoid pedestrians. More advanced systems integrate RL with computer vision, sensor fusion, and predictive modeling to interpret complex scenes and anticipate other road users' behavior. Multi-agent RL has also been explored to coordinate fleets of AVs, optimizing traffic flow and reducing congestion. As AVs approach widespread deployment, RL continues to play a critical role in enabling adaptive, intelligent driving behaviors.

Conclusion: The Road Ahead for RL in Robotics

Reinforcement learning has already begun to redefine what robots can do—enabling them to learn from experience, adapt to uncertainty, and perform complex tasks with minimal human intervention. From robotic arms mastering dexterous manipulation to legged robots navigating rugged terrain and drones coordinating in swarms, RL has proven to be a powerful engine for robotic intelligence. Yet, we are only scratching the surface of what's possible.

Looking ahead, the future of RL in robotics lies in greater scalability, generalization, and autonomy. Emerging techniques like multi-agent RL will unlock collaborative behaviors among fleets of robots—enabling coordinated logistics, search-and-rescue missions, and intelligent factory automation. Meanwhile, lifelong learning and meta-learning will empower robots to continuously evolve, acquiring new skills without forgetting old ones or needing to start from scratch. The integration of RL with foundation models, natural language understanding, and world models could lead to robots that not only learn from interaction but also reason, plan, and communicate in human-like ways.

As simulation tools become more realistic and sim-to-real transfer more seamless, the barrier between virtual training and real-world deployment will continue to shrink. Combined with safer, more sample-efficient algorithms, this will accelerate the development of robots that can operate reliably in homes, hospitals, factories, and beyond.

In essence, RL is not just teaching robots how to act—it's teaching them how to learn. And in doing so, it's laying the foundation for a new era of intelligent, autonomous machines capable of evolving alongside the world they inhabit.

About the Author

Sakina Syed is a dynamic senior data and AI consultant, AI engineer, and enthusiast with a stellar track record in the computer software industry, including notable tenures at tech giants like Microsoft. As a Microsoft AI engineer and Azure-certified professional, she excels in AI deployment, sales, management, teamwork, and leadership.

With a Bachelor of Science in Neuroscience and Mental Health Studies from the University of Toronto, Sakina has also enriched her expertise with courses in business management and IT. Her passion for writing and traveling adds a unique dimension to her professional persona.

Sakina's enthusiasm for business and the application of business psychology to understand consumer needs is matched by her extensive seven-year experience in customer service. She is adept at interpreting consumer feedback and statistical data, enabling her to identify market requirements with precision. Her strong passion for team management and leadership drives her dedication to delivering exceptional customer experiences and fostering business success.

Email: info@youraiconsultant.ca

Website: www.youraiconsultant.ca

LinkedIn: https://www.linkedin.com/in/sakina-syed/

CHAPTER 32

MARKETING AND AI—HOW MARKETERS CAN USE AI TO ACHIEVE THEIR GOALS

By Hanoz Tabak
Strategic Marketing Leader
Toronto, Canada

The playing field is poised to become a lot more competitive, and businesses that don't deploy AI and data to help them innovate in everything they do will be at a disadvantage.

—Paul Daugherty, chief technology officer, Accenture

Over the past few years, artificial intelligence (AI) and marketing have become increasingly connected. AI has a range of applications in marketing, such as helping companies optimize their campaigns to enabling social media influencers to generate content that builds their brands. Before we continue, let's take a moment to define "marketing." It is a term most of us are familiar with, and from my perspective, it is best defined as the art of telling a story to convey an idea, thought or viewpoint. For example, marketing can involve developing your elevator pitch for a job interview or the selling of a product or service by a company (buying shoes from Nike is an example that a majority of people can relate to globally).

When we watch media and movies, our assumption is that marketing is only about large brands creating ads on TV. However, marketing can also involve a team developing a presentation for the CEO on why they should receive a larger budget for upcoming ad campaigns, or influencers advising their followers to purchase products from brands they sponsor. Marketing may not be your profession, but we all engage in marketing in our professional or personal lives; hence, we are all marketers.

For marketers, there are generally two types of customers—internal and external. External customers are people who are buying your product or service, while internal customers vary according to the team or business. They can be senior executives whose help you need to support a project or provide funding for your campaign. Over the past few years, we have all become aware of the opportunities and challenges that AI presents to both internal and external customers. Some of these challenges include ad campaigns that have misled or provided us with inaccurate information about products or services, and internal presentations developed using AI that contain false data.

Marketers should treat AI like fire. Fire can be used to cook food or provide light in the dark when we can't see. It can also be used to burn down a house. Marketers need to be aware of the issues and use AI positively and ethically. This means ensuring that the data used is accurate and not misleading, and that AI tools are used to enhance the customer experience, not to deceive or manipulate. Here are a few ways marketers can use AI to meet the needs of their customers without compromising their ethics or manipulating their customers.

How Can Data Help You?

One of the challenges marketers face is that there is a lot of data available. This data can include customer data, trends from Google, and data from social media platforms like Reddit. The list is endless, and data can overwhelm all of us. For marketers trying to drive more sales or social media influencers deciding what brands to partner with, data can create confusion about the impact of our activities on our brands. The first step is to understand the purpose of the data and how it can enhance your service or offering.

When people read social media posts or stories about AI, they may encounter terms like "deep learning" and "quantum computing." These terms can be confusing, so we'll start with the basics. One way companies use date is for ad campaigns on search platforms. For example, how can we use Google AdWords (the platform people use to buy ads on Google searches) to better understand what search ads my customers like the most? From there, we can start to explore regression analysis, clustering, classification, and dimensionality reduction techniques. Some examples of these techniques include the use of regression analysis to forecast sales volumes or search volume data in Google search platforms, and clustering to understand how customers engage with different search platforms to learn about their products.

Recently, it was announced that a well-known credit card provider was able to use AI on their first-party and third-party data to predict when someone would divorce, with a probability of over 90%, in the upcoming five to 10 years. AI can help you, but you need to use the data ethically; otherwise, people won't trust you. On a side note, first-party data is the information a company or brand collects through interactions with you, while third-party data is collected through aggregators or other sources. For internal customers, data helps companies decide how to spend their time and which projects are driving the most value. An IT company based in Brazil was able to improve its marketing campaigns by leveraging internal data on which admin tasks employees were spending too much time on and then reducing these tasks to help with optimization while improving employee morale.

Before we continue, as the term "data" is used often, here is my viewpoint of the term. When people hear "data" in the news, they think of data breaches or the dark web. Data doesn't focus only on an individual; instead, it provides insights into what a group of people are doing. There are multiple companies, tools, and other resources that can assist you as well—even AI can provide recommendations.

Personalization and AI: Why Should You Care?

As the business landscape becomes more competitive and unpredictable, customers expect companies to deliver better experiences, such as personalization and seamless integration. Brands or entrepreneurs can distinguish their products and services through personalization. AI is a valuable tool for personalization, but it's important to note that this should be done in a way that respects customer privacy and data protection laws. To meet the needs of customers, a brand's first step is to understand who their customer is and the story they are trying to tell about their service or product. There are different ways to understand a customer from researching their demographics to surveys on their aspirations. In the past, when people thought about personalization, they thought about brands sending out emails on a customer's birthday with a discount. Customers have moved past this and want more. Think about it from a customer's point of view—why would they give you their information?

Leading brands in the home renovation product industry are now developing in-app messaging during the buying process. For example, if a customer is buying a shelf, then AI provides messaging on tutorials for assembling the shelf correctly as well as information on how much a shelf can hold without breaking. The benefit of this approach is that it helps the customer and conveys to them that the brand is here to help them achieve their goals rather than just upselling or cross-selling products.

Personalization is valuable, but brands must remember that customers are exposed to at least over 100 messages a day from brands they have previously interacted with and remember there are higher numbers of messages each day from brands they haven't interacted with. From a back-end perspective, companies implementing in-app messaging are doing so using AI tools that learn from customer interactions and feedback. In addition, they ask customers if this messaging is helpful or if it should be removed because, ultimately, the company's story is that they are here to help, and AI is a tool designed to make your life better.

Using external customer personalization for internal customers can help employees distinguish themselves in the workplace. In many organizations, employees are utilizing personalization by allowing their managers to provide feedback on projects through AI tools and then personalizing the next phase of the project—for example, progress reports where managers can select which areas they want to focus on for upcoming meetings. This creates a better experience for both the employee and manager, as it reduces the labor-intensive work of consolidating and personalizing feedback, allowing the manager to direct the employee on where they should put their focus.

Marketers Should Use AI to Make Life Easier

Whether you are a marketing leader managing a team of over 50 employees or an independent business owner who uses marketing to sell their products, there is an opportunity for AI to make your life easier. For all of us, there are things we excel at and things we struggle with. For example, you may have a strong creative background and dislike project management. In this case, you can utilize AI to assist in creating project management tools and tracking requirements. Or perhaps your demanding boss consistently asks for reports that you don't have time to develop. You can use AI to help build a repository of data and then allow your boss to submit requests based on their needs. Reports are created, and you simply need to verify the information. What is the story you are trying to tell your internal customer?

The goal is to focus on how AI can meet your needs and distinguish your brand or service in a competitive marketplace. Below are some additional examples from my personal experience.

- *Researching brands and competitors*: AI tools can help with competitor analysis. I have used AI tools like

Gemini to compare insurance brand websites and better understand which products or solutions companies focus on and how they differ. This analysis includes how each company communicates with prospective customers and how it ranks on search engines like Google.

- *Developing code or schemas to improve our website from an SEO (search engine optimization) perspective*: I haven't had the opportunity to take a course on HTML, but by using AI tools, I can generate code to start discussions with developers on improving web pages. This has helped me learn how to develop and adjust the code or schemas based on my requirements.

- *Improving email content*: Since we all send and receive too many emails, we may need help with what to say or how to express ourselves effectively. AI has been beneficial in rephrasing sentences to enhance the tone of the email and generate ideas on what I may have missed.

The Balance Between AI and Marketing

Let's be honest: AI can be daunting and overwhelming, regardless of who you are. We have all discussed the topic with our friends, relatives, or coworkers, as it is new to most of us. In the world of marketing, there is an opportunity for us to utilize AI to assist customers, both internally and externally. Some of the opportunities include maximizing data, leveraging personalization, and using AI to help us manage tasks and deliver the best experiences. However, we must also be mindful of the potential risks, such as data privacy concerns and the ethical implications of AI-driven decision-making. We must treat AI like fire; it can help us, or it can lead to negative consequences that burn everyone.

About the Author

Hanoz Tabak is a strategic marketing leader and team coach based in Toronto, Canada. He has over 20 years of experience in marketing, focusing on driving ROI for brands and enhancing customer experiences across various industries. His track record includes executing award-winning marketing programs across 20 countries, including tactics in AI, social media, search engine marketing, search engine optimization, digital advertising, programmatic marketing, demand generation, and email marketing. His leadership extends to guiding teams in implementing diverse marketing strategies, acquisition and retention programs, thought-leadership campaigns, and loyalty tactics throughout the digital consumer life cycle.

Hanoz is also recognized for his thought leadership in the marketing community. He has been featured in podcasts and webinars discussing the transformative impact of artificial intelligence on digital marketing strategies. In addition, he has also taught marketing analytics courses at the Schulich School of Business and is passionate about volunteering in the community with organizations like Heaven on the Queensway and ACCESS Community Capital Fund.

LinkedIn: https://www.linkedin.com/in/hanoztabak/

CHAPTER 33

THE DAWN OF CLOUD-NATIVE CUSTOMER ENGAGEMENT: PERSPECTIVES ON CCAAS AND THE AI-DRIVEN REVOLUTION

By Daniel Jonathan Valik
VP Product Management, CCaaS and AI Services
Seattle, Washington

By Hardik Modi
Director Product Management, CCaaS Services
San Francisco, California

In this chapter, we will look into the topic of artificial intelligence (AI) and machine learning within the customer care and contact center industry. It is estimated that billions of customer support interactions occur every day worldwide. How does AI help to solve for better customer care? How is it helping customers and agents, and what transformation do we see in the broader CCaaS market (contact center as a service) as AI is evolving and revolutionizing this space? However, before we dive into AI-enabled customer care and its use cases (and real-world applications), let's deep dive first a bit more into cloud-enabled CCaaS.

The Shifting Sands of Customer Service

What is a contact center? It's one central place where all the customers, like you and me, can "contact" the company to get help, ask questions, or just talk to them. Think of it this way: In the old days, you could only call on a telephone. That was a "call center." But now you can contact customer care in lots of ways. For example, you can:

- *Call* an agent on the phone.
- *Email* them a message.
- *Chat* with them by typing in a box on their website.
- Send them a message on *social media* like Instagram or Facebook.

Because there are so many ways to get in touch, it's called a "contact center." Its main job is to be the friendly, helpful brain of the company that listens to and helps all its customers.

Probably all of us have called customer support services at some point in our lives, whether it's to sign up for a new utility service or to cancel a subscription. It's a ritual that has been going on since the invention of the telephone. In our modern days, it's very often the case that we leverage chat or voice (phone) support to reach an agent with the hope of having a seamless and frustration-free experience. However, you most likely have experienced the opposite and stayed on the phone, getting routed from one agent to another and then never called back, or even worse, lost connection before starting over the entire process again. Customer support has been around for a very long time, and it's surprising that so little has changed. We still want to connect, get help, and have a solution fast, but in so many cases, it's the opposite experience. This is exactly where cloud contact centers and AI play an important role.

For decades, the contact center served as the primary interface between businesses and their customers. Rooted in premises-based telephony systems, these centers were often characterized by their rigidity, high operational costs, and limited flexibility. Agents operated from fixed locations, reliant on disparate systems, and customer interactions were largely confined to voice calls, often resulting in fragmented experiences. These systems could only be afforded by large corporations who had the money and manpower to maintain them.

However, the rapid acceleration of digital transformation, coupled with evolving customer expectations, has fundamentally reshaped this landscape. Today, the traditional "contact center" is rapidly transitioning into a dynamic, interconnected, and intelligent "customer engagement center." At the heart of this metamorphosis lies Contact Center as a Service (CCaaS)—a cloud-native paradigm that is not merely an incremental upgrade but a revolutionary leap forward. This chapter will explore the profound impact of modern customer care services, the underlying forces driving this industry-wide transformation, the pivotal role of artificial intelligence and machine learning, and how innovative players like UJET, in strategic partnership with Google, are uniquely positioned to lead this new era of customer engagement.

Why CCaaS Matters: From "Old" to "New" with Cloud and AI

The migration from on-premise contact center infrastructure to CCaaS is not simply a technological preference, it's a strategic imperative driven by compelling business advantages for nearly any modern business. Traditional contact centers demand substantial capital expenditure (CapEx) for hardware, software licenses, and ongoing maintenance. CCaaS, by contrast, operates on a subscription-based (OpEx) model, eliminating large upfront investments. Businesses pay only for what they use, significantly reducing total cost of ownership and reallocating resources more efficiently from IT infrastructure to customer-centric initiatives. Furthermore, automatic updates and maintenance are handled by the provider, ensuring access to the latest features and security patches without internal IT burden. This allows even a small enterprise to give their customers best-in-class service and access to knowledgeable agents that can answer their every question.

One of CCaaS's most compelling attributes is its inherent scalability. Businesses can rapidly scale their operations up or down to meet fluctuating demand—whether it's a seasonal spike in inquiries, a sudden marketing campaign, or a quieter period. This agility allows organizations to optimize staffing levels, avoid over-provisioning, and quickly adapt to changing market conditions. Moreover, the cloud-based nature of CCaaS facilitates remote and hybrid work models, enabling agents to operate from anywhere with an internet connection, providing resilience and expanding the talent pool.

Enhanced Customer Experience (CX)

Modern customers expect seamless, consistent, and personalized interactions across their preferred channels. CCaaS delivers a true omnichannel experience, unifying customer interactions across digital channels, voice, email, chat, SMS, social media, and even video into a single-agent interface. This centralized view of the customer journey empowers agents with complete context, leading to faster, more accurate, and personalized resolutions. Features like intelligent routing connect customers to the most appropriate agent, while real-time analytics provide insights to continuously improve service quality.

An empowered and efficient agent workforce is critical for superior customer service. CCaaS platforms often feature intuitive user interfaces that reduce the learning curve for new agents. By automating repetitive tasks and providing comprehensive customer information at their fingertips, CCaaS streamlines workflows, reduces agent friction, and allows human agents to focus on more complex, empathetic interactions. This contributes to higher agent satisfaction, reduced burnout, and improved retention rates.

Cloud-native CCaaS platforms are built with open APIs (application programming interfaces), allowing for effortless integration with existing enterprise systems such as customer relationship management (CRM) platforms (e.g., Salesforce, HubSpot, ServiceNow), enterprise resource planning (ERP), and workforce management (WFM) tools. This interoperability ensures that customer data flows freely across the organization, providing a holistic view of each customer and eliminating data silos. This unified data environment is crucial for delivering contextualized experiences and for driving advanced analytics.

Leading CCaaS providers invest heavily in enterprise-grade security measures, often surpassing what individual businesses can achieve on-premise. This includes robust data encryption, multi-factor authentication, disaster recovery protocols, and adherence to global compliance standards (e.g., GDPR, HIPAA, PCI DSS). Migrating to CCaaS offloads the significant burden of maintaining these complex security frameworks from the enterprise, allowing them to focus on their core business.

The Great Transformation: What Innovation Is Happening with AI, and What's the Industry Landscape?

The shift to CCaaS represents a seismic change, compelling both incumbent and emerging vendors to rapidly innovate or risk obsolescence. The contact center market, valued at approximately $45 billion in 2025 and projected to reach up to $60 billion by 2030, is characterized by dynamic shifts and intense competition. If we count in all other areas around CCaaS, including Conversational AI, Generative AI, Workforce Management and CRM integrations, we are looking into an over $100 billion business opportunity for CCaaS vendors already in 2026.

The Inevitable Shift to Cloud-Native

Legacy on-premise contact center providers face the daunting challenge of re-architecting their decades-old solutions for the cloud. Many are attempting to transition their existing customer base to cloud-hosted versions, while others are developing entirely new cloud-native platforms from the ground up, often through acquisitions of smaller, agile CCaaS innovators. This transition is not merely about hosting software in the cloud; it requires a fundamental shift in architecture (e.g., microservices), deployment, and operational models. Vendors failing to make this genuine cloud-native leap risk being outmaneuvered by born-in-the-cloud competitors.

Market Consolidation and Innovation

The CCaaS landscape is witnessing significant consolidation as larger players acquire specialized technologies (e.g., AI, Workforce Management, CRM Analytics) or smaller CCaaS providers to bolster their portfolios. Simultaneously, innovation is rampant, particularly around AI, omnichannel capabilities, and agent empowerment tools. Vendors are continuously releasing new features to differentiate themselves in a crowded market. Gartner's 2025 "Voice of the Customer" for CCaaS highlights the competitive intensity, with players like Cisco and Genesys leading in customer choice, while AWS, Five9, and others maintain strong market presences.

From Monolithic Systems to Agile Platforms

The industry is moving away from sprawling, monolithic applications towards more agile, modular, and API-driven platforms. This allows businesses to select and integrate best-of-breed solutions, creating a highly customized and flexible customer engagement stack tailored to their specific requirements. This architectural shift accelerates innovation cycles and improves overall system resilience.

The Emergence of Artificial Intelligence and Machine Learning in Customer Care—The Brains Behind Modern Customer Care

Perhaps the most transformative force impacting the CCaaS revolution is the rapid advancement and pervasive integration of artificial intelligence (AI) and machine learning (ML). AI is no longer a futuristic concept; it is the fundamental engine driving next-generation customer care, moving beyond simple automation to enable truly intelligent and personalized interactions. The market anticipates a surge in mobile messaging traffic to CCaaS contact centers, largely driven by the implementation of generative AI.

Revolutionizing Efficiency with AI

AI automates routine inquiries, deflecting simple questions through intelligent virtual agents (IVAs) and chatbots, freeing human agents to focus on complex, high-value interactions. Intelligent routing, powered by AI, analyzes customer intent and sentiment in real time to direct inquiries to the most appropriate agent or department, significantly reducing transfer rates and resolution times. Predictive analytics, driven by ML, can anticipate customer needs and even potential churn, allowing for proactive outreach.

Enhancing Personalization and Context

AI enables unprecedented levels of personalization. Through sentiment analysis, AI can gauge a customer's emotional state during an interaction (voice or text), allowing agents to adapt their tone and approach. By analyzing past interactions and customer data (often pulled seamlessly from integrated CRMs), AI provides agents with a 360-degree view of the customer, ensuring relevant and contextualized support. Generative AI, with models like Google's Gemini, Microsoft's Copilot, or Amazon's AI Services (Lex and Alexa), can create human-like responses, summarize lengthy conversations, and even draft follow-up emails, making interactions more fluid and efficient.

Empowering and Augmenting Human Agents with AI

Far from replacing human agents, AI acts as a powerful copilot. Agent assist tools provide real-time guidance, suggesting optimal responses, accessing knowledge base articles, and even coaching agents on tone and empathy. AI-powered knowledge management systems ensure agents have immediate access to accurate information, reducing research time. By automating repetitive tasks and providing intelligent support, AI significantly reduces agent stress and burnout, leading to higher job satisfaction and improved retention. This human-AI partnership allows agents to focus on their unique human capabilities: empathy, complex problem-solving, and building rapport.

24/7 Self-Service and Proactive Support

AI-powered chatbots and IVRs provide instant, round-the-clock support, handling a vast volume of customer queries without human intervention. This enables customers to self-serve at their convenience, improving satisfaction and reducing operational costs. Furthermore, predictive AI can identify potential issues before they arise, allowing businesses to proactively reach out to customers with solutions, transforming reactive service into proactive engagement. Agentic AI is emerging, capable of independently managing and resolving complex tasks, interpreting high-level goals, and interacting with various tools like APIs and databases to automate common troubleshooting workflows.

UJET's Strategy and Competitive Differentiators: An AI-powered Vision with Google

In this rapidly evolving CCaaS landscape, UJET has carved out a distinct and powerful position, fundamentally rethinking customer engagement for the modern, web and smartphone-centric consumer. Its strategy is anchored in a deep, native integration of AI and a unique, foundational partnership with Google Cloud. UJET provides two flavors of their CCaaS service – the UJET (Direct) Contact Center as a Service- and Google Contact Center AI Platform (CCAIP) offering. UJETand Google provide a mobile, web and AI-first approach that is truly shaping the modern ways of customer care and friction-less experiences for customers.

UJET was built from the ground up to address the demands of today's mobile-first world. Recognizing that smartphones are the primary communication device for most consumers, UJET's platform optimizes the customer journey for mobile and web interactions, enabling features like in-app support, secure photo and video sharing, virtual assistant enabled payment methods and biometric authentication for effortless verification. This mobile-centric approach provides a richer context and a more convenient experience for the customer. Critically, UJET is an AI-first platform, meaning AI and ML capabilities are not bolted on but are intrinsically woven into the core architecture, driving every interaction and operational insight.

UJET leverages the world-class infrastructure of Google Cloud Platform (GCP), providing enterprise-grade security, global scalability, and exceptional reliability. Operating on Google Cloud's robust infrastructure, often across multiple availability zones, ensures high availability and resilience. This foundational security and reliability are paramount for enterprises handling sensitive customer data and critical communications.

UJET's key differentiator here is its modern architecture compared to traditional Contact Center platforms and the deep and exclusive partnership with Google Cloud. As the foundational platform for **Google Cloud's Contact Center AI Platform (CCAIP)** service, UJET is evolving customer care use cases with cutting edge innovation powered by Google Cloud Service portfolio to deliver CCAIP to customers worldwide. This unique collaboration allows UJET to deliver a fully integrated, AI-first, and user-first CCaaS solution that is managed, delivered, and supported by Google Cloud itself.

Other differentiators for UJET are the deep integration with Customer Relationship Management services such as ServiceNow, HubSpot or Salesforce, including API's for customization and programmability of many components and use cases. Also analytics and strong reporting functionality is embedded into the UJET and Google platform to give customers everything they need to operate a contact center most efficiently without any 3rd party services. Furthermore, this partnership provides unparalleled access to Google's cutting-edge AI and ML capabilities, including advanced natural language processing (NLP), sentiment analysis, and, crucially, access to Google's latest generative AI models like Gemini. This ensures that UJET's platform is always at the forefront of AI innovation, offering highly intelligent self-service, real-time agent assistance, and sophisticated conversational insights.

Most Common AI Use Cases and What This Means for Customers Interacting with Artificial Intelligence to Solve Their Specific Problems

After learning more about the CCaaS space and its transformation through AI, let's dive a little deeper into the use cases and applications of AI we are seeing currently in this industry and with most vendor solutions, including UJET and Google.

Intelligent Virtual Agents (IVAs) and Chatbots

These AI-powered entities handle a vast volume of simple, frequently asked questions, deflecting them from human agents. Recent advancements, especially with generative AI, allow for more natural, human-like, and empathetic conversations.

- *Intelligent routing*: Beyond basic skill-based routing, AI analyzes customer intent, sentiment, and historical data in real time to direct inquiries to the best available resource— be it a human agent with specific expertise, a specialized bot, or a relevant self-service option. This significantly reduces transfer rates and resolution times.

- *Automated quality management (AQM)*: AI can audit 100% of calls and messages, identifying risks, compliance issues, and coaching opportunities almost instantly, a task impossible for humans at scale.

- *After-call work (ACW) automation*: AI tools can automatically summarize conversations, update CRM records, and draft follow-up communications, reducing post-interaction wrap-up time for agents.

- *Agent coaching and assistance*: AI can provide agents with real-time feedback and coaching, helping them to improve their performance and develop their skills. Furthermore, it could assist the agent in real time with what to ask, what responses to give, what information or solutions to offer and could also provide input on historical events with existing or recurring customers.

- *Agent well-being*: Being a contact center agent is a stressful job. AI can be used to detect agent stress levels and prescribe action, like giving the agent a break or rerouting calls to another agent based on their call-handling performance, voice tone, and language.

AI-Powered Personalization and Context with Data-Driven Insights

AI enables unprecedented levels of personalization by leveraging the rich data collected by CCaaS platforms. Good examples are:

- *Sentiment and emotion analysis*: AI can gauge a customer's emotional state (voice or text), allowing agents and IVAs to adapt their tone and approach dynamically, fostering more empathetic interactions.

- *Customer journey tracking*: By analyzing past interactions and integrating seamlessly with CRM data, AI provides agents with a real-time, 360-degree view of the customer, ensuring relevant and contextualized support.

- *Predictive analytics*: ML algorithms analyze historical data to identify trends and forecast customer needs or even potential churn before they arise, enabling proactive engagement and tailored offers.

- *Hyper-personalization*: AI can go beyond basic personalization, using real-time conversation analysis to feed agents prompts based on a customer's unique experience or emotional needs.

Human Agents with AI

Far from replacing human agents, AI acts as a powerful copilot and force multiplier, enhancing their capabilities in the most common use cases like:

- *Real-time agent assist*: AI tools provide live guidance, suggesting optimal responses, pulling relevant knowledge base articles, or even coaching agents on tone and empathy during an active conversation.

- *Knowledge management*: AI-powered knowledge systems ensure agents have immediate access to accurate, up-to-date information, reducing research time and improving first-contact resolution rates.

- *Training and onboarding*: AI can identify agent skill gaps, recommend personalized training modules, and analyze agent performance to offer targeted coaching, leading to continuous improvement.

- *Automate simple tasks*: By automating repetitive and low-value tasks, AI frees human agents to focus on complex, emotionally charged issues where their unique human empathy and judgment are most valuable, contributing to higher job satisfaction and retention.

Fazit, the Future of Customer Engagement

The journey for the AI-enabled contact center has just begun. The transformation from the traditional contact center to the cloud-native customer engagement center is not just an upgrade, it's a fundamental reimagining of how businesses interact with their customers. CCaaS has emerged as the essential technological backbone for this transformation, delivering unparalleled scalability, flexibility, cost efficiency, the ability to unify disparate communication channels, and, most importantly, the infusion of artificial intelligence in every aspect of customer experience.

The pervasive integration of artificial intelligence and machine learning is accelerating this evolution, empowering both customers through intelligent self-service and agents through sophisticated assistance. AI is transforming customer service from a cost center into a strategic lever for enhanced customer experience, operational efficiency, and ultimately, business growth. This transformation might resolve and lead into fully automated and AI-powered customer care hubs and knowledge centers in the near future; however, for now, we are definitely seeing the first phase of transformation to add AI into cloud contact centers, leveraging AI in countless use cases and semi-automating the overall customer care experience. Perhaps very soon, we will not talk or chat with a human expert but entirely chat with an AI agent across all imaginable topics, from technology, financial, and healthcare support to specific product questions when you, the customer, are purchasing something in any online or physical store. The future of AI helping us has just begun.

In this dynamic environment, a commitment to innovation and cloud-native progress is absolutely paramount. Companies like UJET (or others), through their unique collaboration with Google Cloud, exemplify the future of customer engagement. Combining UJET's mobile-first, AI-first platform with Google's enterprise-grade cloud infrastructure and leading AI capabilities is creating secure, intelligent, and seamless customer experiences that will define the competitive advantage for businesses well into the future. The era of the intelligent, cloud-native customer engagement center is not just on the horizon; it is here, and it is reshaping the very essence of customer care.

About the Authors

Daniel Jonathan Valik is a distinguished author and a seasoned product executive with over 24 years of global expertise at industry leaders including Google-allied UJET, Cisco, Microsoft, RingCentral, Vonage, VMWare, and AWS. A strategic visionary now based in Seattle, he has consistently driven innovation in cloud computing, communications services, data analytics, and generative AI/ML, with a deep focus on augmenting cloud communications with artificial intelligence for next-generation contact centers, unified communications, and communications platforms as a service.

With his portfolio of successfully launched groundbreaking products like Webex Journey Data Service, Microsoft Office 365 Services, Skype for Business and Microsoft Teams Developer Services, Vonage AI Services, Daniel brings unparalleled practical insight. A prolific author of 12 book projects on business and technology, and a renowned global speaker, his work is deeply informed by his extensive international experience, including living and working in Shenzhen-China, Singapore, and Hong Kong. Daniel also has a double master's degree in strategic management and is currently pursuing a study for AI and machine learning at Massachusetts Institute of Technology (MIT). In his personal life, he enjoys spending time with his family, plays musical instruments, and studies martial arts for many years. He is always a curious mind, driven by creative projects and ideas to build, improve, and create something special.

Email: Daniel.Valik@UJET.CX, Daniel@Valik.onmicrosoft.com

Website: www.UJET.CX

LinkedIn: https://www.linkedin.com/in/danielvalik/

Hardik Modi is a dynamic product management leader with a proven track record of launching impactful "zero to one" initiatives within large, matrixed organizations. Based in the San Francisco Bay Area, Hardik excels at recruiting and mentoring product managers, building high-performing product teams, and delivering cutting-edge AI/ML and data experiences for a diverse range of users, including end users, administrators, partners, and executives.

Hardik's expertise spans the entire product lifecycle, from strategic vision to successful execution. He spent over six years at Cisco, leading the Human Agent Insights and Analytics team. At Genesys he led predictive routing solutions using machine learning, generating significant revenue and evangelizing AI products. He also held key product roles at Lattice Engines, specializing in integrated data services and analytics engineering. As director of product management at ujet.cx, he is leading an AI-first approach in contact center as a service (CCaaS) platforms.

Academically, Hardik has continually invested in his growth through Stanford Continuing Studies, focusing on statistics for AI, machine learning, and data science, as well as product management. He holds certifications in generative AI and also is a recognized innovator, holding a patent for a "system and method for predictive routing based on a customer journey patience."

Email: hardike@gmail.com

Website: www.UJET.CX

LinkedIn: https://www.linkedin.com/in/hsmodi/

ALCHEMY IN THE AI ERA: THE MINDSET AND CULTURE THAT TURNS FAILURE INTO INNOVATION

By Jan Wiersma
Entrepreneur, Technologist, and Venture Partner
Amsterdam, Netherlands

Failure is simply the opportunity to begin again,
this time more intelligently.
—Henry Ford

The Alchemist's Promise

In medieval times, the alchemist was a figure of mystique and ambition, dedicated to a single, transformative goal: turning common lead into precious gold. In the digital age, the most valuable alchemy has nothing to do with metallurgy, but everything to do with innovation. The new challenge is to transform the common, inevitable lead of failure into the rare gold of breakthrough success. The rise of artificial intelligence, with its constant stream of failed experiments, flawed data, and costly dead ends, has made this the single most important—and most misunderstood—skill for any modern leader.

Let's be clear about what this new lead and gold look like. The lead is the messy, unglamorous reality of making AI work. It is not a magic wand you wave over your organization. It is the painstaking work of data discovery and cleansing, the endless cycle of experimentation and model tuning, the complexity of understanding which type of AI fits which business problem, and the millions of dollars spent on projects that don't immediately yield a return. It's the promising machine learning model that fails to generalize, the data pipeline that breaks, and the chatbot that confidently hallucinates. It is the raw, heavy, and often discouraging material of technological progress.

The gold, in turn, is the invaluable learning that comes from those failures. It is the resilient, self-correcting system, the deep understanding of a customer's true needs, the innovative product that redefines a market, and the competitive moat that insulates a business. This transformation from lead to gold isn't magic. It is a disciplined, repeatable process that depends entirely on specific human conditions. It is an organizational chemistry that requires three essential ingredients to be present in the right measures:

- The right individual *mindset* to see failure as data.
- The right team *culture* to make failure survivable.
- The right kind of *leader* to skillfully combine them.

This alchemical formula isn't taught in most business schools. Its principles are forged under pressure. For me, the crucible where I first learned these lessons without knowing their academic names was the fireground.

As a volunteer firefighter, I was immersed in an environment where a single mistake could have catastrophic consequences, and where mindset and trust were not soft skills, but literal survival tools. I then carried those raw principles into the business world, spending over 15 years validating and refining them in the trenches as a founder and C-level executive, leading technology companies through the seismic shift to the cloud. My experience proved that this approach was the key to thriving during that disruption. Now, with the accelerated and even more profound disruption of AI, understanding and implementing this alchemical formula isn't just an advantage— it's the only path forward.

Part One: The First Ingredient—The Alchemist's Mindset

The entire alchemical process begins with the quality of your raw materials. In our case, the essential ingredient is the human mindset. Without the right mindset, the lead of failure will always remain lead; it can never be transformed.

Stanford researcher Carol Dweck, in her groundbreaking work, gives us the language to understand this. She identified two fundamental mindsets that govern our response to challenge. A "fixed mindset" is the belief that our abilities are static. This leads to a "performance goal" orientation, where every task is a test to demonstrate one's existing ability. People in this mindset fear failure, avoid feedback that could reveal incompetence, and are motivated by external recognition. In contrast, a "growth mindset" is the belief that abilities can be developed. This fosters a "learning goal" orientation. These individuals are intrinsically motivated by curiosity. They actively seek feedback—even if negative—because they see failure as a non-negotiable part of the learning process.

The iterative and unpredictable nature of AI development makes a fixed, performance-oriented mindset a poison pill for any team. This paralysis stems from a drive to always look competent, where any result short of perfection is seen as a public failure. A team operating this way will be paralyzed by the first model that

JAN WIERSMA

doesn't perform as expected. They will hide mistakes, avoid high-risk experiments that could increase creativity, and see a flawed dataset as an insurmountable obstacle. They will, in essence, give up on the alchemy before it has even begun.

A growth mindset, however, is the fuel for the process. It is the engine of curiosity that drives a data scientist to ask, "Why did this model fail, and what does that teach us about our underlying assumptions?" It is the resilience that allows an engineer to actively solicit harsh feedback on a broken data pipeline, so they can rebuild it correctly the third time. These are the people who see failure as a data point, not a verdict.

As a leader, your job is to become a cultivator of this mindset. You are the chief gardener, responsible for creating the conditions where a growth mindset can flourish. Here's how:

Hire for Curiosity, Not Just Credentials

In interviews, ask candidates to describe a time they failed at a technical project. Don't listen for the solution; listen for how they talk about the learning. Do they take ownership? Do they speak with energy about what they discovered? Ask them to describe a time they received difficult, negative feedback and what they did with it. A candidate who is passionate about what they learned from a failure is more valuable than one who claims to have never failed.

Frame Projects as Expeditions, Not Pass/Fail Tests

When you launch a new AI initiative, frame it as a "learning expedition." Set the primary goal not as "achieve X result," but as "discover the best way to solve Y problem." This intrinsically motivates the team by tapping into their curiosity and reframes failure as a natural part of the discovery process, removing the fear that inhibits true innovation.

Reward the Process, Not Just the Victory

When a team's AI experiment doesn't yield the desired outcome but generates a wealth of new insights, celebrate that learning publicly. Make it clear that the "gold" you value most is not just a successful deployment, but the deep, hard-won knowledge that makes the entire organization smarter. The learning goal-oriented are the alchemists. And they have to be allowed to do their alchemy.

Part Two: The Second Ingredient—The Alchemist's Crucible

Hiring individuals with a growth mindset is only the first step. Alchemy requires a special kind of vessel—a crucible that can withstand intense heat and pressure while the transformation takes place. For a team, that crucible is its culture. Specifically, it is a culture of deep psychological safety.

This concept, pioneered by innovators at places like Xerox PARC under Bob Taylor and later defined by Harvard's Amy Edmondson, is not about being nice or avoiding conflict. It is a shared belief within a team that it is safe to take interpersonal risks. But it's also about something more tactical. A key part of Taylor's genius was acting as a shield, protecting his team from the bureaucracy and politics of the wider organization. He got out of their way, but he also cleared the way for them. This created a bubble of high trust and low friction where people could focus on the work. It's the confidence that you can speak up or admit a mistake without being punished, combined with the freedom to work without being bogged down by corporate overhead.

Without this safety, the alchemy of innovation is impossible. In a culture of fear, the "lead" of failure is hidden, not transformed. A junior data scientist who suspects a dataset is biased will stay silent rather than challenge a senior colleague. A project manager who sees a timeline is unrealistic will agree to it anyway, setting the team up for burnout and failure. A team will choose the safe, incremental path every time, avoiding the ambitious experiments that lead to true breakthroughs. In a low-safety environment, the most valuable data—the data about what isn't working—is suppressed.

As a leader, your most important job is to be the builder and guardian of this crucible. You must actively engineer a culture of psychological safety. Here's how:

Model Vulnerability First.

Safety starts at the top. Begin a high-stakes meeting by admitting something you don't know or were wrong about. Use phrases like "My initial thinking on this was flawed," or "I'm not sure I have the answer here, what are your thoughts?" When you take the first risk, you make it safe for others to do the same.

Run Blameless Post-Mortems.

After every significant project or sprint, conduct a review with one primary rule: Focus on the system, not the person. Instead of asking, "Whose fault was the model's poor performance?" ask, "What can we learn from this result, and what in our process can we improve?" This turns every outcome, good or bad, into an opportunity for collective learning.

Explicitly Reward Candor and Courage.

When a team member points out a flaw in your plan or raises an unpopular but valid concern, thank them publicly. Say, "I really appreciate you challenging that assumption. That's exactly the kind of thinking we need." You are not just rewarding that individual; you are sending a powerful signal to the entire organization about what behavior is valued.

Act as an "Organizational Shield."

Your team is facing pressure from timelines, budgets, and other departments. Part of your job is to absorb as much of that external pressure as possible. You run interference, handle the political battles, and translate bureaucratic requests into clear, actionable tasks, so your team can stay focused on the innovative work only they can do.

Part Three: The Master Alchemist—From Project Manager to Portfolio Manager

We now have our ingredients: a team with a growth mindset and a crucible of psychological safety. The final step is the active work of the master alchemist, who skillfully combines them. This requires a radical shift in thinking: You must stop acting like a traditional project manager and start acting like your company's most important venture capitalist.

Traditional project management is designed to minimize failure and deliver a predictable result on time and on budget. For early-stage AI innovation, this is precisely the wrong model. It guarantees incrementalism. The alchemist, instead, manages a "portfolio of failure," designed to maximize learning. This portfolio has three tiers:

Seed Stage—Many Small Bets

This is where you place dozens of low-cost, fast experiments. Failure is the expected outcome for most. The only goal is to learn as much as possible, as quickly as possible, about new ideas, datasets, and approaches.

Series A—Fewer Medium Bets

The most promising ideas from the seed stage graduate to this tier. They receive more resources and time to build functional prototypes. The goal is to prove technical feasibility and de-risk the concept.

Growth Stage—Very Few Big Bets

A tiny number of validated concepts from the previous stage receive significant funding and organizational support to be scaled into a core product or feature. Failure at this stage is to be avoided, because the idea has been systematically de-risked through the earlier stages of managed failure.

This portfolio approach is more than an operational tactic; it is the engine that builds your company's most critical strategic asset in the AI era. This isn't a new theory I've developed for AI; it's the same battle-tested playbook that separated the winners from the losers during the shift to cloud computing. The move from monolithic applications to cloud-native designs was not just a technical change; it was a cultural one that demanded this exact approach. Small, empowered teams had to experiment, fail, and learn in rapid cycles.

The difference is speed and impact. During the cloud transition, a company's ability to learn from its experiments—what I call its "resilience ratio"—was a leading indicator of market leadership over several years. As a private equity advisor and investor, I can tell you that savvy investors are now looking for this ratio as a primary metric. With AI, where innovation cycles are compressed from years into months, a high resilience ratio has become the single most critical predictor of a company's very survival.

Your blameless post-mortems are no longer just team meetings; they are your deal-flow reviews where you analyze the ROI on failed bets to decide which ideas to fund next. The alchemist's job is to manage this portfolio to maximize the company's resilience ratio, thereby turning cultural strength—your team's mindset and safety—into a hard, defensible strategic asset that investors will value.

Conclusion: The Alchemical Formula for Thriving

The journey to thrive in the AI universe is not, ultimately, a technical one. It is a strategic leadership challenge. The formula is clear: (A

Team with a Growth Mindset + A Culture of Psychological Safety) x A Portfolio Approach = A High Resilience Ratio

This equation reveals the true work of a modern leader. Your mandate is not to pick winners or to avoid failure. It is to build a system that profits from its losses. We learned a version of this lesson when we moved to the cloud, but the pace was more forgiving. The AI revolution offers no such luxury. You must move fast, and you must build a culture that can learn even faster.

Stop managing AI projects and start managing your company's innovation capital. Create a portfolio of bets, celebrate the learning that comes from the ones that go bust, and double down on the ones that show promise. That is the secret to alchemy, and it is the key to thriving in the age of AI.

About the Author

Jan Wiersma operates at the intersection of strategic leadership and deep technical expertise, architecting not just technology but entire business strategies. With a unique perspective forged as a CEO, COO, CPO, and CTO, he translates complex challenges in cloud, AI, and edge computing into clear pathways for growth. His career includes building one of the world's largest statistical machine translation services, developing conversational AI for real-time transcription, and shaping the AI & cloud strategy for major corporations. Currently, as an investor & entrepreneur, he is actively working with multiple AI & LLM-based startups. A visiting university lecturer and VC founding partner, Jan is at the forefront of the AI revolution, guiding the next generation of resilient tech companies.

Email: jan@wiersma.com

Website: www.janwiersma.com

LinkedIn: https://www.linkedin.com/in/janwiersma/

DID YOU ENJOY THIS BOOK?

If you enjoyed reading this book, you can help by suggesting it to someone else you think might like it, and **please leave a positive review** wherever you purchased it. This does a lot in helping others find the book. We thank you in advance for taking a few moments to do this.

THANK YOU

You might also like other Thin Leaf Press titles:

The AI Advantage: Thriving Within Civilization's Next Big Disruption

The AI Revolution: Thriving Within Civilization's Next Big Disruption

The AI Mindset: Thriving Within Civilization's Next Big Disruption

AI: Work Smarter and Live Better Within Civilization's Next Big Disruption

Peak Performance: Mindset Tools for Managers

Peak Performance: Mindset Tools for Sales

Peak Performance: Mindset Tools for Leaders

Peak Performance: Mindset Tools for Business

Peak Performance: Mindset Tools for Entrepreneurs

Peak Performance: Mindset Tools for Athletes

The Successful Mind: Tools to Living a Purposeful, Productive, and Happy Life

The Successful Body: Using Fitness, Nutrition, and Mindset to Live Better

The Successful Spirit: Top Performers Share Secrets to a Winning Mindset

Winning Mindset: Elite Strategies for Peak Performance

Winner's Mindset: Peak Performance Strategies for Success

The Life Coach's Tool Kit, Vol. 1

The Life Coach's Tool Kit, Vol. 2

The Life Coach's Tool Kit, Vol. 3

Ordinary to Extraordinary

The Magical Lightness of Being

Explore.